A Conservation Breeding Handbook

By D. Phillip Sponenberg and Carolyn J. Christman

The American Livestock Breeds Conservancy
Pittsboro, North Carolina

The American Livestock Breeds Conservancy
PO Box 477, Pittsboro, North Carolina 27312

A Conservation Breeding Handbook

Printed in the United States of America by Glover Printing, Inc., Raleigh, N.C.
First printing 1995

D. Phillip Sponenberg 1953–
A Conservation Breeding Handbook / by D. Phillip Sponenberg and Carolyn J. Christman
Includes index.
ISBN 1–887316–00–0

Acknowledgements

Don Bixby has made significant contributions to our understanding of the methods of livestock and poultry conservation, including the specific ideas that became this book. R.D. Crawford has illuminated the importance and complexities of poultry genetic conservation. Anne Ayers, Ed Barraclough, Nils Berglund, John Dawes, Cindy Ehrman, Robert Hawes, Jayne Lowry, Ron Shehee, Torsten Sponenberg and Debby Wechsler provided fresh perspective, useful information, and editorial polishing.

A Conservation Breeding Handbook was supported by grants from CS Fund, Rare Breeds Survival Foundation, and Nan Yarnall. The book was also made possible by gifts from the members of the American Livestock Breeds Conservancy.

Design by Debby Wechsler, with design consultation by David Ashton and Company. Original art for the cover and chapter headings by Meg Kratz.

This book is dedicated to livestock and poultry breeders around the world, in whose hands the genetic treasure rests.

Preface

The American Livestock Breeds Conservancy

The American Livestock Breeds Conservancy (ALBC) is a nonprofit organization which protects genetic diversity in American livestock and poultry through the conservation of nearly 100 endangered breeds of cattle, donkeys, goats, horses, sheep, swine, chickens, ducks, geese, and turkeys.

The American Livestock Breeds Conservancy was founded in New England in 1977 as the American Minor Breeds Conservancy (AMBC). Agricultural historians, while seeking authentic period livestock for interpretive programs at Old Sturbridge Village and other historic sites, discovered that many heritage breeds were nearly extinct. The historians joined with farmers and scientists to form the AMBC with its mission the conservation of endangered breeds. The organization moved to Pittsboro, North Carolina, in 1985. Its name was changed to the American Livestock Breeds Conservancy in 1993 to reflect the growing threat to all livestock and poultry breeds and a widening of conservation activities.

The American Livestock Breeds Conservancy is the only livestock and poultry breed conservation organization in the United States. ALBC does research on breed status and characteristics in order to determine conservation priorities. Programs include the operation of a gene bank; blood typing and DNA fingerprinting for breed characterization; rescues of threatened populations; and the development of genetic recovery breeding protocols. ALBC educates the public and policy makers about the importance of conserving genetic resources and provides technical support for conservation breeding, registry operation, and livestock use to individuals, breed associations, and agriculture organizations. This *Handbook* is an outgrowth of ALBC's technical support programs and is designed to make information on conservation breeding more widely available.

The American Livestock Breeds Conservancy was chartered in Vermont and is a federally designated 501(c)3 nonprofit organization. ALBC is supported by member dues, grants, contracts, sales of publications, and gifts from individuals. You are invited to become a member of the American Livestock Breeds Conservancy. For information, write to ALBC, PO Box 477, Pittsboro, NC 27312 USA.

Contents

Section II. Establishing a Conservation Herd or Flock

Section III. Conservation Programs

List of Figures

Introduction: The Significance of Livestock Breeding

Throughout the long history of agriculture, millions of individuals around the world have dedicated themselves to caring for and breeding domestic animals. Over centuries, people selected from among their stocks those animals which approached a closely held picture of perfection. From these efforts, an array of distinct breeds of animals was created, maintained, and used.

Livestock breeders of today inherit extraordinary genetic wealth in the form of distinct breeds of domestic animals. In the past, it was certain that every generation of breeders would serve for a time as steward for this treasure. There would be some changes at each step, but in general a complete package of genetic material would pass from one generation of breeders to the next. This continuity was an essential characteristic of livestock breeding.

Times have changed, however, and the very traditions of animal breeding are now threatened. Industrialization of agriculture and the urbanization of society have meant that fewer people raise animals. In North America alone, nearly 100 breeds of livestock and poultry are in danger of extinction. Furthermore, similarity among the more popular breeds is increasing; those which were once distinct have become less so.

As a result, future generations may not receive the genetic treasure we have inherited. The genetic diversity essential to selection of animals, environmental adaptation, and maintaining agricultural opportunity may be lost. There is no time to lose. Genetic diversity in the livestock and poultry species—as represented by rare breeds—must be conserved and protected if agriculture is to have a secure future.

The work of conservation rests mainly on individuals, but its success depends on communities of breeders. When many people participate, the result is a decentralization of stocks, which insures against the sudden loss of a single herd or flock to disease or predation. Multiple herds also insure against the insidious losses that can occur gradually if stocks are held as a single group and subjected to a misguided or poorly planned selection program. A network of several herds, each independently managed and selected, is the best way to assure that conservation occurs.

This *Handbook* was written for people who want to become conservation breeders. The book first discusses general aspects of animal breeding as related to conservation, then proceeds to the nuts and bolts of choosing a species, a breed, and individual animals. It continues with a description of the challenging issues of managing small populations for maintenance of genetic viability. There is no single, simple protocol for saving all endangered breeds, but there are skills, principles, and tools that are characteristic of all effective conservation programs.

Some experts worry that individuals cannot be trusted with the heavy responsibility of genetic conservation. Quite the contrary is true. Individual breeders have always been stewards of genetic diversity, and the stewardship practiced by many individuals is the only way to protect this treasure for the future. The *Handbook* is designed to equip breeders, individually and in groups, for this stewardship.

Using This Book

Novices may want to begin with information in Section II on animal husbandry and the choice of a species and breed. The List of Resources provides access to information about each species. A plan for choosing appropriate animals and the numbers to be raised will be a useful basis for understanding the theoretical material in Sections I and III.

Those breeders who are ready to devise a philosophy and conservation plan for their own herds and flocks may begin at Section I for background information and then proceed to Section III for specific breeding protocols. Section II will be of use in the selection of breeds or species for additional conservation projects.

The concepts in this book are often illustrated using cattle as an example, but they apply to all livestock and poultry species. Breeders will find it useful to understand the applications to all species even if their work is focused on only one or two kinds of animals.

Section I.

The Tools of Conservation Breeding

Milking Devon cow

Chapter 1. The Breed Concept: What Is a "Breed"?

"Breed" is an important concept, and its definition must be appreciated before purebred breeding or conservation has much meaning. The best definition of a breed in the genetic sense is based on a concept put forward by Juliet Clutton-Brock: a breed is a group of animals selected to have a uniform appearance that distinguishes them from other groups of animals within the same species. When mated together, members of a breed consistently reproduce this same type.

The creation of breeds took place over a long period of time, and the process produced genetic consistency. Each breed, while containing some diversity within, is uniform and predictable, having a unique combination of genetic traits. Stefan Adalsteinsson refers to this as the "genetic heritage of survival," meaning that each breed's history is embedded in its genetic makeup.

This is why *breed* conservation is important. The conservation of many distinct breeds is the best way to protect genetic diversity within each domestic species. The species themselves (such as cattle, chickens, sheep, etc.) are not endangered, but the extinction of breeds within the species can cause a species to lose some of its characteristics, adaptability, and utility.

The American Livestock Breeds Conservancy has as its mission the conservation of endangered livestock and poultry breeds in the United States. While this work is usually described as genetic conservation, it goes beyond the protection of individual genes. There are approximately 100,000 genes in the genome of each domestic species, and most of these have multiple variations. These variations can be housed in many different combinations. The specific combinations that characterize breeds took millennia to form. Breed conservation is, therefore, more accurately termed genomic conservation, since it protects both genes *and* gene combinations from loss.

Some have proposed that genetic conservation can be accomplished by mixing all of the genes of many breeds into a crossbred soup. Theoretically, specific genes and combinations could then be recovered if and when they were needed. This method would be time consuming and expensive, if indeed it is possible. In contrast, the conservation of living breeds proven for specific tasks and environments protects genetic resources in an accessible form. The combinations already present as breeds are the genetic resources most easily studied, used, and conserved.

Breeds are the genetic resources most easily studied, used, and conserved.

As a final note, some livestock and poultry populations that are named, recognized, and registered as breeds do not really fit the genetic definition of a breed. In most of these, crossbreeding has eliminated the genetic consistency and predictability that is characteristic of a true breed. Such populations are not of conservation interest.

Types of Breeds in North America

Four types of livestock and poultry breeds occur in North America: landraces, standardized breeds, industrial stocks, and feral breeds. These differ according to their history and selection, the relative level of uniformity versus variability, and the presence or absence of a formal structure to organize and promote them. Figure 1.1 provides an illustration. While the majority of North American breeds (and endangered breeds) are standardized, a discussion of the different types is useful in explaining how breeds were developed and how each should be conserved.

Landrace breeds

Landraces (not to be confused with the Swedish Landrace swine breed) are local populations of animals that are consistent enough to be considered breeds, but are more variable in appearance than are standardized breeds. They also lack the formal definition and organizational structure that is typical of standardized breeds.

Individual landrace breeds are unique due to founder effect, isolation, and environmental adaptation. Founder effect refers to the accidents of history that led to introduction of certain types of animals to new areas. These "founders" are the sole genetic base of descendant populations. For example, Texas Longhorn, Florida Cracker, and Piney Woods cattle of the U.S. (as well as the sister criollo breeds of Latin America) were all founded by a few hundred Spanish cattle brought to the New World beginning in 1493, and this fact forever defines these populations. Coupled with founder effect, and difficult to separate from it, is the action of isolation. Even given identical founders, populations that are isolated from one another will eventually diverge over time. The differences today between Texas Longhorn, Cracker, and Piney Woods cattle are an example of the divergence that occurs due to many centuries of isolation.

Piney Woods cattle are forever defined by their Spanish foundation.

A combination of natural and human selection has shaped the evolution of landrace breeds. Natural selection and geographical isolation have created genetic consistency and adaptation to the local environment. Traits such as parasite and disease resistance, reproductive efficiency, and longevity have also resulted. Human selection is of somewhat less importance. In fact, human selection in one part of a landrace population may be counteracted by different human selection in another part. Color is one example. The Holt line of Piney Woods cattle is usually white park or colorsided roan with black ears, while the Conway line is red and white in various combinations. Both individual herds have lost some color variants (gaining uniformity of color in the process), but the landrace breed has not.

Landrace breeds generally lack the formal support of a breed association, and they survive as distinct populations due to geographic and cultural isolation. If communication among breeders increases, and a network of breeders is organized, the landrace may benefit by greater geographical distribution and more secure numerical status. This process can, however, result in selection for greater uniformity across the population and diminish the presence of some of the original variants. If, instead, there is careful cultivation of the diversity within breed parameters, the genetic integrity of the landrace can be protected even as it becomes a standardized breed.

Standardized breeds

Historically, most livestock breeds began as landraces and then became standardized. As breeders organized, they agreed upon a description (or "standard") of the breed and then began to select their animals toward this ideal. In practice, this meant greater uniformity in appearance and performance. Genetic diversity may have been reduced, but predictability was gained. The Texas Longhorn is a good example of a landrace that is becoming a standardized breed.

Standardized breeds are what most people think of when they consider purebred livestock. Human selection has played the primary role in the development of standardized breeds, though natural selection has sometimes played a part as well. The breed standard defines the breed, and it is this criterion by which individual animals are evaluated. Breeding to a standard emphasizes a relatively narrow range of variation, usually less than the variation found in a landrace. Genetic isolation of the standardized breed is thus established by breeding practices rather than by the geographic or cultural isolation typical of landraces. For example, most standardized breeds limit inclusion to those animals with two registered parents. This serves to isolate standardized breeds genetically. It also allows breeds to function as isolated reproductive units even though herds may be widely separated geographically.

The American Mammoth Jack is a highly developed standardized breed.

Conservation of standardized breeds (as well as landraces) requires that the purity of the breed be protected from crossbreeding. The diversity within the breed's genetic parameters must also be conserved. The two forces of predictability and variability tug against one another, and striking a balance between them is a goal of breed conservation.

It is often the case that certain color patterns or conformational traits have become hallmarks of standardized breeds. These traits are sometimes so distinctive that even casual observers can identify members of the breed. As a result, many people, even those in the livestock industry, have come to confuse a single, distinctive color or trait with breed status. Ironically, some of these traits—such as white belts, white faces, or polledness—can be readily imparted to crossbred offspring, so that they are actually among the least accurate indicators of purebred status. A group of animals must breed true to a distinguishing type to be a true breed.

White Leghorns, an industrial breed, are the most numerous egg laying chickens around the world.

Industrial stocks

Industrial stocks originate from a small portion of a standardized breed (or a cross between two or more standardized breeds) that has subsequently been selected for maximum performance of a single task in a highly controlled environment. Examples are layer chickens, broiler chickens, white broad-breasted turkeys, and most commercial swine. Almost all of our chicken and eggs, much of our pork, and a growing proportion of milk and other animal products come from industrial stocks. "Industrial" refers to the highly organized production systems under which the animals are raised, and the general term "stocks" is used because industrial animals may be breeds or crosses between breeds.

These stocks are a superb example of limited genetic variability and supreme predictability, having been defined by relatively few founders and long isolated from other breeds. Human selection plays the major role and natural selection is minimal.

Industrial stocks (such as Cornish-Rock broiler chickens) are sometimes *raised* by individuals, but they are seldom if ever *bred* by individuals. Breeding stock is instead closely held by a few multinational corporations. The selection of these stocks is very complicated, and the environment in which they function is likewise highly developed and organized. As a result, there is little role for the individual breeder in the maintenance or conservation of industrial stocks. This is not to say that genetic loss does not occur in industrial stocks; it does. Bloodlines may become extinct following company mergers or changes in priority. The conservation of these genetic resources, however, is primarily the responsibility of the corporations which control them.

Feral breeds

Feral animals are those domestic stocks which have escaped or been released from human ownership and returned to the wild. Feral populations which reproduce in the wild have thus reverted from human selection to natural selection. Most feral populations across North America are not breeds in the genetic sense; continual

Santa Cruz Island sheep are a feral breed now being conserved in domestic flocks.

introduction of new animals has eliminated the genetic consistency characteristic of a true breed.

There are, however, a small number of feral populations which are considered breeds, as they have been created by distinct founders and isolated for a long period of time. The genetic significance of feral breeds includes environmental adaptations and relict traits not found in improved breeds. If the founders are now extinct, feral breeds assume even greater importance. Ossabaw Island swine and Santa Cruz sheep are among the feral breeds found in North America.

Management and conservation of feral breeds is a challenge. The most significant feral breeds have generally evolved in areas of geographical isolation (such as islands) which may also be home to genetically significant native animals and plants. It is often the case that feral populations must be removed or extirpated to protect the survival of the other organisms. In these situations, it is important that the feral stocks be evaluated to determine their genetic significance, so that those which warrant conservation can be protected elsewhere. The irony, however, is clear: how can environmental adaptation be conserved outside the habitat? Conservation of feral breeds done *ex-situ* (outside of the habitat) requires an intricate breeding protocol to maintain genetic variation and reduce additional human selection.

Figure 1.1 Types of livestock and poultry breeds in North America

Type of breed	Uniformity and predictability	Variability
Landrace	Moderate for traits of survival	High (usually) for traits of appearance
Standardized	Moderate to high	Moderate to low
Industrial	High	Low
Feral	Depends upon founders and selection pressures	Depends upon founders and selection pressures

Cotswold sheep

Chapter 2. Developing a Philosophy of Breeding

A basic and guiding philosophy is the single most critical component of any breeding program. A breeder's first task is the development of a specific purpose in breeding animals. This may seem obvious, but it is very often overlooked, with the result that breeding is done with little progress toward any goal. Perhaps even worse than an aimless approach is a breeding program that chases after the fads dictated by the show ring or by other people. Such a program is constantly acquiring, discarding, and thus wasting genetic resources in a futile attempt to catch up to the current leaders in the breed.

A philosophy of breeding combines the elements of genetics, selection, and husbandry to accomplish a goal. If a philosophy is in place, every animal and every mating can be evaluated relative to the goal, giving purpose and identity to the herd or flock.

This chapter considers the philosophies of conservation and improvement. An understanding of the methods and goals of each of these will enable breeders to develop their own philosophies of animal breeding.

The Philosophy of Conservation Breeding

The philosophy of conservation breeding has as its objective the management of a breed as a genetic resource for the future. The goal is to maintain a breed's original genetic identity, including its production potential, by using selection criteria consistent with the breed's history. In addition, selection is made for physical

The conservation philosphy focuses on the group, such as this Myotonic goat herd.

soundness. This means that sound animals of breed type will be used in reproduction, with careful attention to maintain all existing bloodlines. (An exception may occur in the case of extremely rare breeds or strains, where it is not always possible to cull all unsound or blemished individuals until numbers have been increased.)

The conservation philosophy focuses on animals as a group rather than as individuals. Attention is placed upon a breed, or on a herd or flock of a breed. The genetic variation present in a single herd or flock will be managed so that this smaller group serves the breed's long term genetic needs.

The most dramatic management change required by the conservation philosophy is the selection and use of a relatively larger number of individuals (particularly males) for breeding. In addition, a sound animal somewhat lacking in appearance or performance but representing less common genetic variants may be used for mating instead of one which is superior in appearance or performance but whose genes are already well represented in the herd or flock.

The popular interpretation of the word "conserve" is to save things exactly as they are, but this is not exactly the case. Living animals continue to evolve, and conservation allows these changes to occur. Conservation of livestock and poultry includes their continued use and evolution in the habitat of agriculture. Those breeds that are treated as museum pieces will surely become extinct.

Preservation, in contrast, does mean keeping things exactly as they are. Animal genetic preservation is accomplished through cryopreservation, which is the freezing of semen, embryos, and other genetic materials for the future. This technology can be used to ensure the survival of all male bloodlines in a herd, flock or breed, and it can provide the opportunity to reintroduce bloodlines that become lost from the population. Preservation is not, however, complete in itself. A breed kept *only* in frozen storage is, ironically, at greater risk of loss because its characteristics and utility will be forgotten. Effective genetic conservation programs for domestic breeds must include living populations.

The Philosophy of Improvement Breeding

The philosophy of improvement breeding has as its goal the increase of performance for specific production or conformation traits in a breed, herd, or flock. The idea of making animals more useful, productive, and even beautiful has always been the driving force in the breeding of domestic species. To understand the philosophy of improvement breeding, however, it is necessary to go beyond its broadest objectives and look at how breeding for improvement is practiced today.

Within the improvement breeding philosophy are two distinct and sometimes opposing objectives: the increase of uniformity in a population; and the production of outstanding individuals. The first objective guides most production animal breeding programs, since herd or flock uniformity, especially uniformity of performance, is highly desirable. The second objective, with its concentration on individual performance, is more common in horse breeding and in the production of show stock of any species.

Most improvement programs are focused on performance for a single production trait, or a very few production traits. These include rate of weight gain in beef cattle, volume of milk production in dairy cattle, and speed in race horses. The disadvantage of this approach is its reliance on very narrow selection criteria. If *every* animal lacking superb performance for the trait is culled, this can lead to loss of variation for other traits. If *every* animal that shows superb performance for the trait is bred, then other, undesirable characteristics may be perpetuated.

Dutch Belted cattle are an improved breed selected for milk production in forage based systems.

The trait of body size has often been the focus of improvement programs, and production breeds in every livestock species have been selected for very large stature. This selection has resulted in biological problems, including longer reproductive intervals, difficulty of birth (dystocia), delayed maturity, unsound feet and legs, and shorter lifespan. In miniature animals, selection has been for small size, since the smallest animals can bring the highest prices. Small size, however, can be caused by congenital and hereditary problems, so miniaturization can inadvertently select for these defects and leave animals unfit. Breeding for any extreme cannot be considered beneficial and is instead often destructive to the genetic health of a population.

In a more sound approach to improvement breeding, selection is based on performance for several production traits rather than a single trait or only a few traits. Such a program results in more balanced performance animals and fewer genetic losses. Dairy cattle breeders, for example, could follow the example of the Dutch Belted, a rugged, sound dairy breed whose strengths include excellent production on forage, reproductive efficiency, maternal ability, and milk composition.

Conservation and Improvement

The philosophy of improvement need not conflict with the philosophy of genetic conservation, and it is always desirable to include both in a breeding program. This is especially true for breeds which historically have been shaped by an improvement philosophy. In these breeds, improvement should continue to be a part of breed conservation. Examples include the Oxford sheep as a crossing sire in meat production systems, the Leicester Longwool sheep as a crossing breed for meat and wool production, and the American Mammoth Jackstock ass breed, used to produce elite mules for work and riding.

Jacob sheep in America and Britain have diverged due to selection differences.

Improvement and conservation conflict only when improvement represents a departure from the historic uses and characteristics of a breed. Such transformation usually implies loss of the breed's genetic heritage. Jacob sheep are an example. The Jacob sheep in America reflects its origins as a relatively small and primitive breed, but the British Jacob has been improved to fit the modern marketplace and is now much larger and smoother than in the past. Though the breed name is the same, the original breed type of the British Jacob has largely been lost.

Pig breeding poses a particularly difficult challenge in the light of conservation and improvement strategies. Most rare pig breeds are fatter than the current pork market desires. These breeds were skipped over when intense selection for leanness and against fat was used in the development of modern commercial types. To select today's rare breeds for leanness might better assure their survival, but the cost would be loss of original genetic type. Again, the names could be saved, but the hogs would be different.

These examples illustrate the fact that no single philosophy is correct for every breeding program. Individual breeders must develop their own philosophies based on the genetic status and heritage of their breeds. As breeders' experience grows, their philosophies will evolve. What remains constant is the importance of philosophy. An understanding of the concepts of conservation and improvement, along with conscious reflection on the goals of a specific breeding program, will greatly increase the chances of accomplishment.

Bronze turkey

Chapter 3. The Concept and Practice of Selection

All breeds have been developed and are continually shaped through the process of selection. *Webster's Dictionary* defines selection as the act or process of choosing some individuals from a group based on fitness or preference. In the case of livestock and poultry breeding, selection results in the survival and propagation of some individuals but not of others, with the result that the inherited traits of the survivors are perpetuated. Selection is at work any time that animals in a group do not contribute equally to the genetic pool of the next generation.

Selection acts on the genetic variation found in a population. Most of the population's characteristics will have many possible expressions, or variants. Individual animals may express a characteristic in different ways, or to different degrees. Selection favors some of these and penalizes others. In this way, the direction of the population is determined. Selection will slowly whittle away at variation and create a more uniform final product, but the original range of variation and the rate of selection will determine the end result.

Domestic breeds have been shaped by both natural selection and human selection. Natural selection favors survival. The animals best fitted to the environment will survive longer and produce more offspring, passing along more of their genes to the next generation. The unsound and unfit generally do not survive or reproduce as effectively, and their traits will gradually diminish in the population.

Piney Woods cattle, for example, have been shaped by five hundred years of natural selection in the hot, humid climate of the Southeast U.S. The result has been relatively

Navajo-Churro sheep have been crafted by both the environment and human breeders.

small animals with high fertility, short calving intervals, and a long lifespan. All characteristics in a population, however, are not affected by natural selection; those which do not affect survivability may remain variable. The Piney Woods cattle breed, for example, has kept a wide variety of coat colors over the past five hundred years, since natural selection in its environment gave no survival advantage for any color over another.

Human selection is a much more recent activity. Humans have bred animals for several thousand years. It has only been in the last two centuries, however, that an understanding of reproduction and genetics has enabled human breeders to standardize breeds. The late nineteenth century was the pinnacle of breed development in every domestic species. Twentieth century technologies have greatly expanded the power of human selection, and breeding has become more oriented toward maximizing production performance. Breeders can greatly increase the number of offspring that individual animals produce and can even use cloning to replicate single individuals many times over. Breeds that could not survive in nature —such as the broad-breasted turkey—have been developed and maintained by human breeders, using the ability to operate production environments which are independent of nature.

It is important to understand that each breed of domestic animals has a distinct selection history. Conservation programs should be consistent with the history of the breed being conserved. Some breeds, such as the Santa Cruz Island sheep, were created primarily by natural selection. Others, such as the Leicester Longwool sheep, have resulted from tightly focused human selection. Most breeds fall between these two extremes, having been shaped by both nature and human breeders. The Navajo-Churro sheep of the American Southwest has evolved due to natural selection and also the selection practiced by human breeders for strong, naturally colored fiber. This breed illustrates the profound relationship that exists between human societies, domestic animals, and the environments they share.

Positive and Negative Selection

Selection has two aspects, the positive and the negative. Positive selection is the choice of animals *for* reproduction. Negative selection is the choice *against* the use of animals for reproduction. Positive and negative selection are usually directed at a few specific characteristics, but selection acts on the entire animal. All of its genetic traits (whether under selection or not) are either perpetuated or lost.

Culling is the selection against an animal for reproduction and its removal from the breeding group. Culls can be sold, neutered, or used for meat, depending on why they have been culled. Some animals will be culled after a breeding career because they are no longer needed in the herd or flock. This is particularly the case in conservation breeding programs, where more males are used over a shorter time than is typical of ordinary breeding programs. Breeding stock no longer useful in the herd of origin can be sold to other breeders. In this case, the animals are being culled from a herd or flock but not from the breed.

Selection is usually directed at a few specific characteristics but acts on the entire animal.

Sound individuals with performance value may be culled from reproduction by neutering. This is particularly common in horse and donkey management, where neutered males have added performance value as recreation, companion, and guardian animals. Neutering is also used, though less often, in the management of cattle (as steers are used for oxen) and in sheep (as wethers may be used to produce wool). Neutered animals have been culled from the breed, though they may continue to have value as representatives for the breed.

Animals which are unsound or deviate from breed type should not be used as breeding stock. Rather, they should be culled from the breed as well as the herd or flock by being neutered or used for meat.

Selection Intensity

The intensity of selection refers to the relative proportion of animals which are used for breeding versus those which are culled. In intense selection, a relatively small number of animals are bred and a relatively large number of animals are culled. Genetic variation may be rapidly reduced as the genetics of a few individuals are established throughout a population and much genetic material is discarded.

Unimproved Bronze turkeys retain the ability to mate naturally.

Intense *natural* selection generally occurs in the wake of dramatic environmental change, such as famine or introduction of predators, causing most of a population to be lost. Intense *human* selection has the same effect, though for different reasons. In dairy cattle breeding, computerized production projections may determine the single bull to be used for breeding and the thousands of others to be culled.

The opposite of intense selection is relaxed selection, in which a greater proportion of animals are retained for breeding and relatively few are culled. Relaxed selection retains variation in most of the characteristics present in a population and the change that occurs does so slowly. This type of selection is found most often in breeds which are extensively managed and relatively unimproved.

An example of intense human selection is the introduction of the broad-breasted characteristic in turkeys. The Broad-Breasted turkey originated in England and was imported to North America in the 1920s. The commercial advantage of the broad-breasted characteristic was significant, and, through intense selection, all commercial turkeys became broad-breasted. Many of the show varieties also acquired this characteristic.

By the middle of this century, continued selection for breast meat had succeeded to such a degree that the toms could no longer get close enough to the hens to mate. The extra white meat on the breast (so attractive commercially) now makes it necessary for all commercial turkeys to be bred using artificial insemination. The successful improvement of domestic turkeys has come with significant biological cost—loss of the genetic potential for natural mating. The individual breeder's ability to reproduce turkeys has thus been reduced or even eliminated. This very important commercial bird is biologically unfit, and it is now completely dependent on humans for its survival.

The turkey is not alone. Intense selection without regard for fitness and utility has led to problems in several modern breeds. When sheep were intensively selected for wool coverage, for example, the result was expected to be heavy fleeces. In actuality, this selection produced "wool blind" sheep with poor survivability. Intense selection for rapidly growing, large framed sheep has resulted in a high incidence of the spider lamb defect, a hereditary disorder. Selection for big, muscular Quarter Horses perched on small, dainty feet has resulted in a variety of foot problems in this breed.

Intense selection without regard for fitness and utility has led to problems in several modern breeds.

To be successful, breeders must take into account the impact of selection on the whole animal, and on all future generations. Natural selection has been at work for thousands of years to mold each species. In the attempt to rapidly change a particular trait, we risk upsetting the inherent genetic balance that is present. This is why animal breeding still remains an art as much as a science.

Uses of Selection

Selection in standardized breeds

Standardized breeds have been developed from the careful crafting of past breeders. Conservation breeders of today work within that tradition, selecting their stock according to the breed's standard and its history, with attention focused on production parameters and environmental adaptations. Breeders must maintain the genetic variation within the breed as well, and breed standards should be written to describe the entire range of characteristics appropriate for a breed rather than a single, narrow ideal. Both the predictability and the diversity within breed parameters are important.

This is not to say that any and all diversity within a breed is valuable. If a chestnut Cleveland Bay horse is born from registered parents, the chestnut color does represent "diversity" in a literal sense, since all other members of the breed are bay. Such an individual would be culled since it does not meet the standard of the breed. The goal of selection is to conserve the diversity within the breed's heritage, and animals which deviate from this are culled.

Soundness should be a selection priority in standardized breeds, and it should include reproductive soundness, maternal ability, hardiness, and general vigor. Animals must be able to show appropriate levels of performance under appropriate management.

Selection in landrace breeds

Landrace breeds are appropriately managed under "extensive husbandry," that is, out of doors, on a forage-based diet, with minimal veterinary input. This type of husbandry is not particularly difficult. The challenge of landrace breed conservation, however, is the use of selection techniques to maintain the characteristics which have been shaped by natural selection—such as disease and parasite resistance, maternal ability, fertility, climate adaptation, and production on very low inputs. How does the human breeder select for these complex attributes, which are not readily visible?

While landraces are known for their hardiness, there is a tendency to tout qualities such as parasite and disease resistance without allowing selection for them to occur. Without selection, these traits will gradually be lost. Instead, nature must continue to take its course. The herd or flock must be challenged by the environment so that resistance can be demonstrated and selection for resistance can be made. Exposure to parasites, for example, is the only practical way that individual animals can demonstrate their degree of parasite resistance. Careful evaluation of individuals, plus the use of parasite counts in feces, will allow the breeder to select resistant individuals for reproduction. If the goal is to maintain disease resistance, then diseases must be allowed to occur, with minimal intervention through vaccination or treatment programs.

Gulf Coast Native sheep must continue to be selected for parasite and disease resistance.

This is not to recommend that treatment be withheld from individuals which become parasite-ridden or ill. These animals should treated, but—if the goal is to select for resistance—they must then be culled from the breeding population. Nor is this to say that effective conservation breeding eliminates the need for good husbandry and veterinary care. Instead, the idea is to match the husbandry and care to the specific breed and to use selection to maintain the original array of characteristics.

Selection for parasite or disease resistance can be most effectively accomplished by saving those animals that keep the bloom of health in the face of a challenge, and culling those that show the adverse effects. Prior to selection, the herd will contain some unthrifty individuals, but resistance (or lack thereof) must be demonstrated in order to maintain this characteristic in a population.

Selection in feral breeds

Breeders who work with captive populations of feral breeds find themselves conserving a gene pool selected by the harsh hand of natural selection. It is practically impossible to replicate such selection pressures under human management. The best course is to maintain the existing genetic variation within a population, particularly all of the founder bloodlines, with minimal human selection. Selection can also be done to maintain resistance to parasites or disease, but caution is advised. The specific challenges present in new, managed environments are not necessarily the same as those found in the isolated, original habitat. Breeders should use great care when challenging these stocks with parasites or diseases. In the case of very rare breeds, numbers should be increased first and a challenge done only later, and then only on a portion of the population.

Selection should be used to maintain the origninal array of characteristics in a breed.

Even with a careful conservation program, a feral breed will gradually be transformed under human management, and at some point is no longer truly feral. Conservation breeding can, however, ensure that a good portion of the genetic diversity represented by feral breeds is not lost.

Responses to Selection

A very important and complicated detail of animal breeding is that the genetic control of various traits is different, and each trait responds differently to selection. Keith Hammond of the Animal Genetic Resources Group of the Food and Agriculture Organization of the United Nations has emphasized an important phenomenon that has long been known but has been largely ignored: traits related to production (including rapid growth, meat conformation, milk production, and the like) respond relatively quickly to selection pressure while traits of adaptability (such as reproductive fitness, climate tolerance, and parasite resistance) usually respond relatively slowly to selection pressure.

This has significant implications for breed selection and conservation, both in the

Tamworth pigs retain the adaptability necessary for production on pasture.

United States and around the world. Those "native" breeds which are the best adapted to adverse environments are generally less productive than other breeds, and it can be tempting to replace or crossbreed them with imported production breeds. Imported animals, however, often fail to thrive in a new environment, much less reproduce, leaving people dependent on continual reimportation of high-cost replacements. Crossbreeding, while it will retain more of the adaptive characteristics and give hybrid vigor as well, is ultimately a dead end unless pure stocks of the native breed are maintained.

More secure long term progress in the quest to improve performance can often be made by selecting *within* a native breed for production characteristics, because these traits respond relatively quickly to selection pressure. This strategy is far more likely to produce long term success, since selection for adaptability within a production breed is practically impossible.

Modern breeders have available the technology and information to make rapid changes in breed populations, including the elimination of large amounts of genetic material. It is essential that conservation breeders understand the goals of breeding and make considered use of the powerful tool of selection.

Cleveland Bay horse

Chapter 4. Breeding Systems

Selection of animals always occurs against a backdrop of breeding systems. These are as important as selection, for they produce the animals upon which future selection will be used. Inbreeding, linebreeding, linecrossing, and crossbreeding are discussed in this chapter. The terms can be emotionally charged for many breeders, but each system can be useful in animal production and conservation programs. Also discussed are random and multi-sire mating, which are management techniques independent of breeding systems.

Inbreeding

Inbreeding is the mating together of animals which are related so that the resulting offspring have one or more ancestors that occur on *both* the sire's side and the dam's side of the pedigree. Inbreeding can vary from extreme (or "close") to less extreme (or "distant"). Examples of close inbreeding would be the mating of full siblings or parent to offspring, while more distant inbreeding would be half-siblings or cousin to cousin.

Inbreeding itself is neither bad nor good. The quality of the result depends entirely on what goes into the mating. Put good animals in, and good animals will uniformly come out. Put in any genetic weaknesses, and weak or deformed animals can result. Importantly, if good-looking animals hiding genetic weaknesses are inbred, then these weaknesses will be exposed. For that reason, inbreeding should always be coupled with a program of selection for excellence in breed characteristics, viability, and fertility, with the strict culling of any animals that show weakness. Inbreeding should be used

The Shorthorn was an early product of inbreeding.

in moderation; a general guideline is to increase the inbreeding coefficient by no more than 5% per generation (see Chapter 8).

Inbreeding has negative connotations for most breeders, but it can be a powerful tool to increase the consistency and uniformity of a population. Inbreeding has been used, either purposefully or accidentally, in the establishment of most breeds. For example, the Florida Cracker, Piney Woods, Texas Longhorn, and the other criollo cattle breeds in the Americas descend from a foundation of perhaps no more than three hundred cattle brought to the New World by the Spanish. Numerous standardized breeds, among them Shorthorn cattle, were shaped by inbreeding in their early days. Indeed, many rare breeds and strains are characterized by some degree of inbreeding, and most of these are fully viable.

The use of inbreeding does require close attention to detail. Some breeders will calculate the inbreeding coefficient in their herds or flocks, and the use of this formula is discussed in Chapter 8. Of more importance is the careful observation of the population of interest. When levels of inbreeding pass certain thresholds, most groups of animals experience "inbreeding depression." This is the decline in reproductive fitness and in the vigor of the young produced. Breeds will vary in their tolerance of inbreeding, and some seem unusually resistant to inbreeding depression. Even so, it is always advisable to keep available an outcross within the same breed to use in case of inbreeding depression, so that the survival of a population does not depend on the continued success of inbreeding.

Linebreeding

Technically, linebreeding is a form of inbreeding that seeks to concentrate the genetic impact of a single excellent individual (rather than several) throughout a population. The goal is to create a group of animals as much like the excellent individual as possible. The most common linebred mating is a half-brother to half-sister, though cousin matings are also considered linebreeding.

The difference between linebreeding and inbreeding is subtle. As one breeder said, "it's linebreeding if it works and inbreeding if it doesn't." This reflects the common assumption that linebreeding is good, inbreeding is bad. Though most of the comments about inbreeding are also true of linebreeding, linebreeding is less extreme, and its benefits can be gained with a lower risk.

Randall Blue Linebacks are a linebred remnant of a cattle type once common in New England.

Linebreeding has been the specific strategy used in the development of many standardized breeds, such as the Morgan horse, in which early matings were designed to concentrate the blood of the stallion Justin Morgan. This strategy has also been used in the development of bloodlines or strains within breeds, such as the Farceur line of American Belgian horses.

The strength of both inbreeding and linebreeding is that they narrow the range of variation, making the resulting population more uniform and therefore more predictable. Purebred animal breeding frequently has this as one of its goals: a uniform, predictable, high quality herd or flock. The weakness of inbreeding and linebreeding is the possible loss of vigor and reproductive performance. To prevent this, rigorous culling should always be a part of an inbreeding or linebreeding program.

Linecrossing

Bloodlines (or strains) are sub-breed groups that are more closely related to one another than they are to the breed as a whole. The existence of distinct bloodlines within a breed has significant benefits for all breeders, among them the opportunity to use the method of linecrossing. Linecrossing is the breeding together of individuals from different bloodlines. It is somewhat like crossbreeding, in that the genetic distance between two distinct lines will create a performance boost in the first generation.

As a result, linecrossing is often used to produce outstanding individuals. This is an especially good tactic for the production of show stock of any species, since linecrossed animals tend to have a lot of bloom and presence. Linecrossing can also be used periodically to increase vigor in a linebred population.

Crossbreeding

Crossbreeding refers to the mating together of different breeds, and it can be a useful and fascinating practice. The initial result of crossing two distantly related breeds is a sometimes spectacular performance boost called hybrid vigor or heterosis.

Crossbreeding is most often used today to get superior production stock. An example is the common practice of crossbreeding Angus and Hereford cattle. The initial offspring, or F1 (for first filial generation) offspring, are uniform "black baldy" cattle that are black with white faces. If these black baldies are bred to each other, however, the results are quite nonuniform: black cattle, red cattle, white-faced blacks, white-faced reds, horns on some, others polled. Colors, horns, body type and performance will all vary between the Angus extreme and the Hereford extreme.

The initial uniformity gained by crossbreeding is due to genetic uniformity. This is based on the fact that the F1 offspring have one Hereford gene and one Angus gene at each gene site. The next generation's variability results from the resegregation of genes into Hereford pairs, Angus pairs, Hereford/Angus pairs, and Angus/Hereford pairs. Multiply that by the hundreds of thousands of genetic sites present in cattle and it is easy to understand just how variable the results can be. While in theory it would be possible to reassemble a purebred Angus or Hereford from the mix, this is practically impossible.

Black baldy cattle are a common commercial cross of the Hereford and Angus breeds.

Though crossbreeding is a one-way street away from purebred parent stock, the resulting genetics can be used in the development of new breeds. Breed crossing in the middle to late 1800s was the basis for almost all poultry breeds developed in America. The Barred Plymouth Rock, for example, was created in the 1860s and 1870s from crosses of the single combed Dominique, Black Cochin (or Java), Black Spanish, Gray Dorking, Brahma, and perhaps others. Many generations of breeding and selection produced the genetic consistency necessary for the new breed to be established.

The success of crossbreeding is dependent upon the existence of genetically distinct parent breeds. If parent breeds become similar to each other, the performance boost available from crossing them will be reduced. This is a problem for the swine industry, for which crossbreeding has been the cornerstone of commercial production but very few genetically distinct breeds remain alive today.

It is worth noting the role of crossbreeding in conservation programs for rare breeds. Males from rare breeds can be bred to females of other breeds in order to produce offspring that combine performance with soundness and other qualities. Dutch Belted bulls, for example, are often bred to Holstein cows and Cleveland Bay stallions to Thoroughbred mares. In each case the goal is offspring marked by hybrid vigor and a useful combination of the parents' characteristics. Such use of crossbreeding can promote a rare breed and demonstrate its utility, attracting new breeders and resources to pure breeding programs.

While the use of males in crossbreeding may serve a rare breed conservation program, females must be bred pure to replenish the rare breed population. Crossbreeding the females is a waste of their genetic value, as well as one of their limited number of breeding seasons. The crossing of Cleveland Bay mares with Thoroughbred stallions and of Texas Longhorn cows with improved beef breed bulls has in each case produced many desirable production animals. In both cases, however, the crossbreeding has hastened the demise of the purebred stock that is necessary to produce the crossbreds. This is very like killing the goose that lays the golden eggs.

Inbred, linebred, and linecrossed animals can all be useful in the conservation of rare breeds such as Suffolk horses.

Which is Best?

The choice of using inbreeding, linebreeding or linecrossing in a herd is an individual choice for the breeder to make. If the goal is an excellent *herd* with high predictability and decreased variability, the choice should include some linebreeding or inbreeding. This generally implies a long term commitment to a specific bloodline within a breed and it can produce excellent long term results. In addition, the animals produced in an inbred or linebred herd are useful to other breeders, largely due to the genetic predictability that these animals offer. This is true within a breed, and true between breeds.

If the goal is the production of excellent *individuals*, then linecrossing may be the approach to take. In the long run, this will produce excellent performance animals, though these individuals will not in their own turn always produce a uniform population of offspring.

The genetic characteristics of linebred, inbred, and linecrossed populations also have consequences in the selection of new individuals added to a breeding program. Consider a situation in which one of two bulls will be added to a herd to provide some genetic distance to a closely related population. One bull is linecrossed, while the other is inbred or linebred. If the quality and appeal of the bulls is similar, it is usual to expect the inbred/linebred bull to outproduce the linecrossed bull. This is because the inbred/linebred animal is more genetically uniform, and he will generally produce offspring better than himself, especially if used, as in this case, for linecrossing in the new herd.

In contrast, the linecrossed bull is already the result of certain favorable combinations and some hybrid vigor. It is likely that with such an individual "what you see" is not necessarily "what you will get" when he is used for breeding. Such a bull will likely produce some offspring like himself, but he will not do this as uniformly as the inbred or linebred bull. This is a good general rule, with some exceptions.

Successful crossbreeding is dependent on access to genetic diversity. This is why commercial producers have a vested interest in breed conservation.

The characteristics of linebreeding and linecrossing also have significance for crossbreeding and breed conservation. Successful crossbreeding is dependent on access to genetic diversity. This is why commercial livestock producers have such a vested interest in breed conservation; the conservation of distinct, pure breeds is the only way to protect the performance advantage to be gained by crossbreeding. If genetic diversity is lost through breed extinction, then this powerful breeding method will be of much less use to breeders of the future.

Random Breeding and Multi-Sire Mating

Random breeding, or random mating, is sometimes confused with breeding systems such as inbreeding, linebreeding, or linecrossing. It is instead a management technique independent of breeding system. Random breeding, in the strict sense, means that replacement animals are selected at random and pair mated without regard to appearance or performance. Random breeding is the specific mating system that is usually assumed to be operational in populations studied by theoretical geneticists in the exploration of the impact of population size on genetic structure.

Random breeding is sometimes used to describe multi-sire matings. In this system, multiple males are allowed to run with a group of females and there is no direct control over the specific matings that take place. Multi-sire matings can be used to good effect in many breeding programs, but they deviate from true random breeding in two ways. The first is that the group of males chosen to mate have generally been selected from a larger group of males that were considered as potential sires. The second is that if all the males are used at the same time in a breeding group, the result will likely be that the most socially dominant male will sire the most offspring, with other more submissive males having less (or no) offspring. In a truly random breeding system the females would be randomly allocated to males, and a distribution of matings would be assured by a single-sire system.

Chickens are often produced using multi-sire mating systems. This is a White Plymouth Rock.

The multi-sire mating system is especially common in chicken and turkey breeding and it will often succeed if care is taken to assure a relatively broad genetic contribution from a number of sires representing different strains, not just a single dominant male. This system can also be used to good advantage for primitive adapted breeds and landraces, since it provides for some continued selection based on environmental adaptation and social interaction.

Section II.

Establishing a Conservation Herd or Flock

Standard donkey

Chapter 5. Choosing a Species

A conservation breeder must first decide which species to raise. The traditional livestock species in North America are cattle, donkeys, horses, sheep, goats, and swine. Also considered to be livestock are the poultry species, which include chickens, ducks, geese, and turkeys. Livestock species less common in North America (such as llamas, alpacas, yaks, and water buffalo) are not specifically addressed in this book, though many of the strategies described can be applied to conservation breeding of these animals.

Each species has a unique combination of characteristics, with specific husbandry needs and production uses. A breeder should choose the species which best fits his or her interests, capabilities, and resources. Conservation is hard work, and part of its reward is the pleasure of raising animals one enjoys. Novices unfamiliar with livestock and poultry should spend some time at local fairs and farms to see which species are most interesting and attractive to them.

A breeder's capabilities are important to consider. What size animals are manageable? What kinds of husbandry are appropriate? It is important to be honest about how much time, energy, and money are available for the care and management of the animals. Chickens, for example, require only a little time but they generally need attention each day. Dairy cattle and dairy goats must be milked at least twice a day while in production. Beef cattle and horses may require less daily care, but there are no maintenance-free animals of any breed or species.

Resource Needs and Management Options

Resource needs for livestock production include feed, forage, buildings, fencing, and water. Of particular interest is the type of forage (grass and browse) available, which may determine which species are appropriate and how many a particular piece of land will support.

Each species eats a slightly to markedly different selection of plants, and there are some important breed variations within each species as well. For example, horses and most breeds of cattle eat grass primarily. Sheep prefer grass and leafy plants. Goats as well as some breeds of sheep and cattle are "browsers," eating brushy plants as well as grass, so these species can be used to clear land and open it up for better grass production. Pigs and poultry have traditionally been kept in orchards where they forage for some of their own food, control pests, and aerate the orchard floor. Pigs also eat fallen fruit, and the apple-flavored pork that results was traditionally regarded as a delicacy. Geese are excellent grazers, keeping pond banks and other wet areas mowed. Well managed grazing and browsing provides both animal food and effective land management, and as a breeder's experience in forage management increases, so will the efficiency of land use.

Poultry can be managed on pasture using movable chicken houses.

Other features of the land are also important. What amount and quality of water is available? Do watershed or other zoning restrictions limit land use? Are predators a problem? What type of fencing is present or desired? An old adage recommends fences be "horse high, bull strong, and hog tight," and a wide variety of fencing options make this possible. What potential markets for animals or products exist, and where are they? An evaluation of the resources available and necessary for different species is a useful process. Local Cooperative Extension agents and other breeders and farmers are often good sources of information.

The management of each species will reflect the capital, labor, and environmental resources of the breeder as well as the characteristics of the species. There are several management options being used in America. The most common is the industrial model, which involves heavy inputs of feed,

These Tamworth pigs are managed under an intensive grazing system to maximize pasture productivity.

energy, veterinary care, housing, and equipment. The objective is maximum production of animals or animal products to bring a high return. This may or may not generate much profit, however, due to the high costs of production.

A different management option is sustainable agriculture, meaning operation of an agricultural enterprise with reduced off-farm inputs. The objective is to have the farm sustain itself. This type of production has much in common with the traditional diversified farm, which included a variety of crops and animal species. In such a system, animal services may be used to replace inputs of fossil fuels and chemicals and thus reduce production costs. Many rare breeds are quite appropriate for sustainable management systems, as they are hardy, forage efficient, and climate adapted. These breeds are also appropriate for systems that fall between the sustainable and industrial extremes.

Within sustainable agriculture, there are several specific animal production methods currently being used. These include "management intensive grazing" (MIG), where animals are rotated through a succession of grazing paddocks; "integrated pest management" (IPM), which is the use of animals, insects, and other biological controls for pests; free-range production of poultry and pigs; and agroforestry, which is the use of poultry and certain livestock for insect and weed control in fruit and nut orchards.

The benefit of sustainable management systems is that while production costs are much lower, the animals often show remarkable health benefits and both animal and plant products may bring a premium price. Organizations such as ATTRA (Appropriate Technology Transfer for Rural Areas) provide information about sustainable agriculture, including livestock management, to commercial producers. These appear in the List of Resources at the end of this book.

The following sections give a brief description of the characteristics and uses of each of the livestock species. Additional information on each species is available from organizations and materials in the List of Resources. Figure 5.1 begins the introduction process by listing the gender names for each species.

Figure 5.1 Nomenclature for each species

Species	Female	Male	Neutered male	Young
Cattle	Heifer (young) Cow	Bull	Steer Ox (if trained)	Calf
Chickens	Pullet (young) Hen	Cockerel (young) Cock Rooster	Capon	Chick
Donkeys	Jenny Jennet	Jack	Gelding	Foal
Goats	Doe Nanny	Buck Billy	Wether	Kid
Geese	Goose	Gander		Gosling
Ducks	Duck	Drake		Duckling
Hogs/Pigs	Gilt (young) Sow	Boar	Barrow	Piglet Pig
Horses	Mare	Stallion	Gelding	Foal Colt (male) Filly (female)
Sheep	Ewe	Ram Buck	Wether	Lamb
Turkeys	Hen	Tom		Poult

Cattle

The dominant feature of cattle is their size. Adult cattle weigh 600 to 3,000 pounds and therefore require the use of a fairly large amount of land. Fencing should be strong and secure. It is recommended that handling facilities (such as chutes or stanchions) be available for restraint in the case of even minor veterinary treatments such as dewormings and vaccinations. Otherwise cattle can harm themselves or their handlers, even if inadvertently.

Most breeds of cattle do well on pasture, and well-managed grazing with distribution of manure can significantly improve grassland. Several breeds of cattle (such as the Devon, Galloway, and Highland) are browsers as well as grazers, and they can be used to clear brushy pest plants to allow for better growth of grass.

Cattle produce both milk and beef, but most breeds are specialized for a single product. Management of dairy cattle is a demanding occupation, with milking required at least twice a day, every day they are in production. The handling and marketing of milk is complicated, requiring specialized facilities and knowledge. As well as the production of milk, dairy cows can be used (in place of artificial feed) to raise calves, and each cow can often raise several calves, depending on her level of production and maternal abilities. Rare breeds of dairy cattle will perform best under grass-based management. Indeed, it is only under systems where forage efficiency and longevity of production have economic value that such breeds will be competitive with Holsteins. All dairy cattle are eventually used for beef, and a breed's meat traits will also affect its value in a dairy herd.

Galloway cattle are adapted to a range of climates.

Beef breeds do not require as much daily attention as dairy cattle and they may do nicely with minimal management. Markets for beef cattle are well established throughout the country, though markets may penalize unusual breeds because of their deviation from the standard commercial profile in such characteristics as size, horns, hair, and so forth. Some rare breeds of cattle may have value as breeding stock above base slaughter price, although this usually depends on an effective marketing program. Custom butchering services are available in some areas for those breeders who want to sell beef directly to consumers, and this is often the best way to gain a good return from the production of grass-fed beef.

Dual purpose cattle breeds were historically selected for good production of both meat and milk, with beef selection predominating for the past few decades. Dual purpose cows produce a lower volume of milk than the specialized dairy breeds, but the milk is richer in solids, making it excellent for cheese manufacture. Health requirements for milk sold for cheese are less stringent than those for milk sold as liquid milk, but the decreasing number of outlets for such sales may limit farmers' opportunities.

Cattle have a nonseasonal reproductive pattern, but most are mated to produce spring or fall calves. Gestation is roughly nine and one-half months, and a calf left with its mother may nurse six to seven months before weaning. Cows produce one calf at a time, twins only rarely. Sexual maturity is reached at about one year, and heifers generally produce their first calf at about two years of age.

Cattle may be pasture-bred using a bull or they may be bred with artificial insemination (A.I.). A.I. technologies are well established for cattle and generally available, important when a breed is widely dispersed or when breeders want to use several different bulls. At the same time, however, extensive use of artificial breeding has been a powerful selector for a few characteristics—such as collectability and freezability of semen from males and obvious estrus in females. Individuals which lack these characteristics do not reproduce well using A.I. and they can therefore be eliminated from a gene pool, with drastic consequences for a breed's diversity.

Piney Woods cattle have historically been used as oxen in the deep South.

Cattle in North America are generally kept in production for five years or less, though their natural lifespan is longer than this. Many of the rare breeds, in fact, live into their teens. In general, cattle are productive throughout their entire lives and only rarely lose fertility due to old age.

Cattle can be trained for draft work, and working cattle are called oxen. Though theoretically any individual can be trained to work, oxen are most often steers (neutered males), which are easier to work with than bulls. Oxen are regionally important in New England and in pockets elsewhere. The breeds and strains of cattle known for draft work can be of great local importance and prestige. Milking Devon cattle, for example, were traditionally valued in New England as athletic, quick, and intelligent oxen, and this played a pivotal role in the breed's conservation.

Horses

Horses and ponies in America are used primarily for recreation as riding and driving animals. They produce no tangible products except offspring, but they do offer a variety of services, including draft power, transportation, and sport. Horses work closely with people, so much of their worth comes from the skills developed through training.

Recreational activities with horses include: cross country riding, driving, endurance competitions, hunting, jumping, polo, and trail riding. In addition, a wide variety of horse show events exhibit the athletic ability, speed or style of individual animals. Horses are often used to manage cattle, and draft breeds can sometimes be more economical than tractors or other farm machinery.

Horse breeds come in a wide variety of sizes, temperaments, and types. The height of horses is traditionally measured at the withers in "hands," with a hand equal to four inches. For example, 14 hands is equal to 56 inches, and 14 and a half hands is equal to 58 inches. Miniature horses can stand less than nine hands (36 inches) and weigh only a few hundred pounds. The average riding horse is about 15 hands and weighs about 1000 pounds. Draft horses may be over 17 hands high and weigh more than a ton. Like cattle, horses are large animals, and size alone makes them potentially dangerous. It is advisable to have confinement facilities for routine or emergency veterinary treatment.

Morgans, developed as a multipurpose breed, are often used for driving.

Like the other livestock species, horses can be sold for meat at commercial auction, and most horses sold at auction are bought for meat. Horse meat is used domestically for pet and zoo animal food or shipped abroad for human consumption. The slaughter markets are

Horses, such as this Florida Cracker, are primarily used for riding.

well organized and are responsible for the base price paid at auction, which forms a price floor throughout the industry.

Horses are managed under a variety of systems. Highly specialized show animals are stabled year round and fed grain and hay. Most pleasure horses, however, will thrive on pasture and hay when given proper shade, windbreaks, and water. Horses are selective grazers, and they can be hard on pasture. Two or more acres of average forage may be required per horse per year in addition to hay during the winter. Skillful management of forage can reduce the amount of land needed. Optimum land management would combine pasture rotation with the sequential or simultaneous grazing of other species (such as goats, cattle or sheep) to eat a wider array of plants.

Horses are widespread across North America. In most areas there are networks of knowledgeable breeders and veterinarians. In addition, horse care and training is described in hundreds of books. The most difficult subject to research is management of pasture for horses, though this information is gradually becoming more available.

Equine reproduction is seasonal, in the spring. Mares have a gestation of eleven to twelve months, so mating and foaling occur at the same time of year. Foals are nursed for four to six months, then weaned before the next foal appears in the spring. Twins are uncommon, and they rarely survive. Horses are generally not used for breeding until they are three years old, although stallions can reach sexual maturity as early as one year of age.

In contrast to the owners of other species, most horse owners are not horse breeders, nor do they intend to become breeders. Instead, people generally use individual horses as athletes, which precludes simultaneous use for reproduction. Economic return on horse breeding is slow, since young stock generally bring much lower prices than do older, trained animals which have required long years of investment. In addition, horse breeding itself is somewhat technical, and few people have the facilities and experience to maintain a breeding stallion. Some stallions are sweethearts and others are truly dangerous; still others can be either, depending on the day. Artificial insemination is available for some breeds, though it does add to the complexity of a breeding program, particularly as mare heat cycles vary in their detectability.

Horses can live up to thirty years or so, although most are considered to be elderly in their twenties. They are consistently productive up to their mid-teen years, both for reproduction and for athletic performance, but afterwards both reproduction and athletic performance is reduced.

People who want to become conservation horse breeders should probably first learn management and reproduction techniques on more common horse breeds. Once the basics are mastered, this experience can be put to valuable use in the many tasks of conserving rare horse breeds—keeping stallions, raising foals, and training and promoting stock.

Donkeys

Donkeys, more accurately called asses, are closely related to horses but are a separate species. Donkeys in America, like horses, are primarily used for recreation.

Mules, the hybrid offspring of horses and donkeys, can be excellent riding animals.

Donkeys range in size from the Miniature, less than thirty-six inches at the withers, to the Mammoth Jack, which stands over 15 hands and weighs over one thousand pounds. The smaller breeds are generally companion animals, while larger ones are used in recreational activities such as riding, driving, and packing. The reputation of donkeys as stubborn is largely an offshoot of their propensity to freeze up and avoid dangerous or unknown situations. If wisely managed, this trait can serve an owner to good advantage. Donkeys are also known for their dislike of dogs and coyotes; because of this trait, they can be effective guardians of sheep, goats, and other small animals. Single females or castrated males are generally used as guard animals, as uncastrated males tend to be too rough and can even threaten their charges.

One of the most important historic uses of donkeys in the United States has been the production of mules. A male donkey (or jack) bred to a female horse (or mare) produces a mule, which is a sterile hybrid combining the horse's size and strength with the toughness of the donkey. The characteristics of the mule

Donkeys are hardy, social animals.

depend on the breeds of its parents, and mules can vary as widely as horses and donkeys. Mules were traditionally used for draft power in agriculture and in the military, though today there is growing interest in the use of lighter weight mules for pleasure riding and driving.

Donkeys evolved in the deserts of the Middle East and North Africa, and their hardiness far exceeds that of most other domestic species. Despite this, regular veterinary and hoof care is necessary, and finding professionals with experience in donkeys can be difficult in some parts of the country. A common management problem in the America is the overfeeding of rich, high protein grain diets, when donkeys do best on forage. Donkey management is very similar to that of horses and they can be kept together in many situations.

The breeding of donkeys is also similar to that of horses, though artificial breeding is not as widely used. The management of male donkeys, or jacks, can be even more difficult than that of stallions, so specialized facilities are desirable. Management of Mammoth Jacks should not be undertaken by anyone without some experience with either horses or other breeds of donkeys.

The gestation of donkeys is twelve to thirteen months, one month longer than that of horses. As a result, jennies usually do not have a foal every year, but will occasionally skip a year in order to allow earlier rebreeding for a spring foal. A three-year-old animal is considered mature, although many are mated younger. Donkeys usually live longer, and sometimes much longer, than horses. Individuals in their forties are not uncommon.

Sheep

Sheep have one of the longest histories of domestication, and they are one of the most useful animals for a diversified farm. Sheep produce meat (usually in the form of lamb), wool, milk, cheese, and other dairy products. In addition, they are good grazers and browsers, able to clear pastures of unwanted brush. As an old Spanish proverb says, “wherever sheep’s feet touch the ground, the land turns to gold.”

Sheep, such as this Tunis, are among the most useful of the domestic species.

Breeds of sheep vary in size and shape, with adult weights ranging from less than 100 pounds to over 400 pounds. Even the large breeds are, however, much smaller than cattle or horses, so they are attractive for small properties. Sheep are kept in groups, or flocks. They have been selected for the flocking instinct and require the security of their own kind. A single sheep will be very unhappy and so will its hapless owner.

Care of sheep need not be complicated, and most breeds are easily managed on pasture. Sheep eat such a variety of grasses, forbs and brush—including invasive pest plants such as leafy spurge, multiflora rose, and honeysuckle—so successfully that they may be used for biological weed control. A run-in shelter for protection against wind and rain is sufficient for sheep, though in most areas they must also be protected from predators with a combination of measures, including fencing, guard animals, and secure housing.

Sheep require regular preventive veterinary care, including deworming and hoof trimming. Most rare breeds are generally healthy, though wooled sheep tend to “hide” very effectively under their fleeces, and the early, treatable states of a problem can be missed. For that reason, the shepherd must develop a keen eye for the condition of the animals. Wooled breeds must be sheared once or twice annually, and shearing is a technically and physically demanding task. In some areas of the country, shearers are difficult to locate and breeders may have to learn this skill as part of their own flock’s management.

Sheep are usually mated in the fall for spring lambs, though some breeds will reproduce year round. The gestation is about five months. Breeds vary in fecundity, though twins are fairly common and some breeds produce even higher multiples. Lambs are weaned at two to three months of age, and sexual maturity is attained at about six or seven months, though there are some breed differences. Sheep are usually productive until eight to ten years of age, though individuals can live into their middle to late teens.

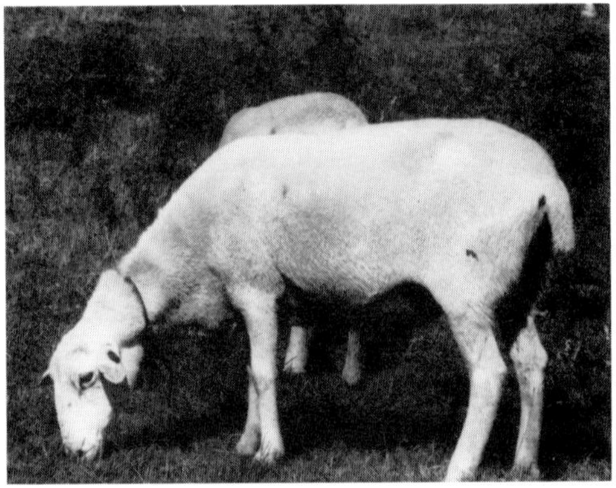

Katahdins are hair sheep, used in meat production.

The meat from sheep is generally sold as lamb. Markets for lamb are unevenly distributed across North America and strongest in New England. The best markets are in cities and other areas of ethnic diversity, though good promotion will sell good lamb (and lamb sausage) anywhere. Sale of meat directly to consumers will bring a higher return than sale of live animals through commercial auctions, where deviation from the standard meat sheep profile is usually penalized.

The "hair sheep" breeds are particularly useful for meat production. They have hair, not fleece, so they do not require shearing. The smaller size of some of these breeds will be penalized at auction, but their hardiness, parasite and disease resistance, fertility and maternal abilities can be significant economic assets. This is particularly true in contrast to commercial sheep production, for which veterinary care has become a tremendous expense. The group of hair sheep breeds in America includes the Barbados, St. Croix, and other Caribbean breeds, which are well adapted to warm climates. Another hair sheep breed is the Katahdin, developed in the 1970s in the U.S. It is a larger framed breed, derived from a blend of Caribbean hair sheep, Wiltshire Horn, Suffolk, and other breeds.

Sheep's milk, richer and less allergenic than cow's milk, is used in cheese, candy and other products. The demand for sheep's milk is increasing in America as consumers have the opportunity to try it. Several breeds of sheep show both the high levels of milk production and quiet disposition necessary for dairying, though the recent importation of specialized East Friesian dairy sheep genetics will probably swamp any other breed's contribution to the North American industry.

Wool production and marketing

Wool breeds produce one of four general types of fiber: fine wool, medium wool, longwool, and carpet wool. Fine wool sheep produce fiber of the smallest diameter. The commercial fine wool breeds, Rambouillets and Merinos, have been selected for uniformity of fleece texture and white color. These breeds are important in the international fiber markets for the production of stockings, underwear and other clothing and textiles. The Shetland, also a fine wool breed, has been selected for a variety of natural colors (in addition to white) and its fleece brings a premium price in the handcraft market.

Medium wool fibers are larger in diameter than fine wools. Most commercial meat sheep produce medium wool, and consequently this is the most common type of wool produced in North America. Medium wools are readily available and are considered a specialty fiber only if the color, texture or some other characteristic is unusual. The Black Welsh Mountain, Dorset Horn, and Jacob are among the rare breeds that produce medium wool.

Longwool breeds, including Cotswold, Leicester Longwool, and Lincoln, produce fiber that is longer and of greater diameter than the medium wools. Longwool is characteristically shiny and lustrous. In addition, naturally colored Cotswold and Lincoln sheep have fleeces in shades of black, silver, and gray. The longwools are very easily spun by hand, and because of their colors and textures they are highly valued by handcrafters.

Leicester Longwool sheep produce long, lustrous wool.

Carpet wools are completely different in type. They combine a long, coarse hairy fiber which is large in diameter with a soft, fine inner fleece. Carpet wools are used for the manufacture of rugs and other textiles where resilience and strength are desired. Relatively few breeds of sheep produce carpet wools; those found in America include the Karakul, Navajo-Churro and Scottish Blackface. These wools are penalized in normal wool marketing channels, but they are actively sought by the handcraft market.

Most wool produced in America is sold into the industry-sponsored "wool pool" program. The wool pool consolidates large batches of wool from a geographic area for sale to manufacturers. Since uniformity of wool pool fiber is highly desirable, already low wool pool prices are usually further discounted for specialty wools which have unusual color, texture, or quality. In contrast, the revival of spinning, weaving, and other fiber crafts has created a demand for specialty wools. This has been of great importance in the conservation of rare breeds of sheep. Properly marketed, the price for specialty wool far exceeds the wool pool price.

Information about wool production and marketing is available from local and regional spinning and weaving organizations, sheep breed associations, and industry publications. Wool marketing is quite variable by region, and each breeder should plan to research and then develop a market for fleece in order to get the premium price that good quality wool should bring.

Goats

Goats have many of the same characteristics as sheep, although they are generally more inquisitive and adventuresome. These qualities make them engaging, and they can become great pets. Goats are highly sociable animals and they get along well with other species. In fact, a single goat should not be kept alone; goats need the companionship of other animals.

The management of goats and sheep is similar, and they are easily kept together. Goats are hardy and generally require little more than routine preventive veterinary care, such as deworming and hoof trimming. Like sheep, goats will do well in a shelter that keeps them dry and draft free, but they must be protected from predators with a combination of fencing, housing, guardian animals, and other measures. At the same time, goats are more athletic than sheep, and it can be a challenge to keep them inside a fence. Goats love to eat browse, so they are excellent for ridding land of brambles, honeysuckle, kudzu, poison ivy, and other fast growing pest plants.

Goats can be trained to pull carts.

In addition to their importance in land improvement, goats are used to produce meat, milk, and fiber. They are also used as companions and in recreation, including work as pack and driving animals. The variety of goat breeds, which range in size from thirty to 300 pounds, gives the goat enthusiast a great deal to choose from.

Markets for goats and their products are not well organized nationally but are growing. A breeder with the ability and time to invest can develop good markets for goats and goat products. Opportunities for meat sales already exist in the South and in ethnic urban markets, and they are growing wherever the diversity of the population is increasing. Meat goats are often sold at auction, though direct access to processing plants and consumers may improve a breeder's financial return.

Goat dairy products are widely used around the world, and several breeds have been specifically selected to produce milk. Dairy goats, like dairy cattle, require milking twice daily. They produce far less quantity than do dairy cows, but this smaller amount can often be to the advantage of a household. Though relatively unfamiliar in North America, goat milk and cheese are increasing in popularity as a health and gourmet food. Goat milk is less allergenic than cow's milk and easier for many people to digest. Distinctive goat cheeses are considered an attractive specialty item.

State regulations set requirements for operation of goat dairies and the sale of dairy products. The sale of fresh, fluid milk is usually regulated more strictly than is the sale of goat cheese. The investments in equipment for dairy enterprises can be substantial, and prospective cheese makers must also invest time in learning this ancient art.

Fiber is produced by Angora and Cashmere goats. The Angora produces mohair fiber, and the Cashmere produces cashmere. (Angora fiber is produced by Angora rabbits.) Fiber producing goats must be sheared, and this can be more difficult than the shearing of sheep since goats tolerate restraint poorly.

These San Clementes demonstrate the adventuresome nature of goats.

Female goats of most breeds are strongly seasonal breeders, producing spring offspring called kids. Some breeds, though, can breed year round. Twins and triplets are common, even more so than among sheep breeds. The gestation is about five months, and kids are usually weaned after a two or three month lactation. Goats mature rapidly, with sexual maturity as early as three to four months old. Goats are productive for eight to ten years, although some may live twice that long.

Though artificial breeding with frozen semen is becoming available, most breeders will need access to a buck to breed their females for the production of kids and milk. Bucks can be aggressive and they do have a distinctive odor which many people (including the neighbors) may find offensive.

Pigs

Pigs, also called hogs or swine, are intelligent and inquisitive large animals, with some weighing over 1,000 pounds. Pigs can be charming as well as useful, and this accounts for the current interest in miniature breeds (80-100 pounds) as pets.

Pigs were once an essential part of the diversified farm, used for meat, fat, leather, and bristles. They were extensively managed, expected to forage for acorns, glean fields after harvest, and consume kitchen, dairy, and brewery waste, as well as windfall fruit in orchards. Today, pigs are kept solely for meat production; they are generally kept confined and eat grain-based feeds. The agricultural services of pigs, such as turning of compost and manure piles and control of Johnson grass and other invasive plants, are not widely utilized, though there is a glint of interest in this potential among practitioners of sustainable agriculture.

The industrialization of swine production this century has had a dramatic impact on swine genetics. Only a few breeds are used commercially, and all of these have been selected for leanness and rapid growth based on high protein feeds and intensive husbandry. The rare breeds are rare today because they do not thrive under confinement or because they are generally fatter than today's market prefers.

The habitat of Gloucester Old Spots and other rare pig breeds is the diversified farm.

As a result of industrialization, outdoor swine production has nearly disappeared, and an entire generation of farmers has missed an opportunity to learn about these methods of husbandry. Only recently is extensive production being tried once again, but it will not necessarily be successful when used with breeds selected for rapid growth in confinement on high grain diets. Farmers interested in raising pigs on pasture would do better to find breeds that retain the qualities of hardiness, initiative, and maternal abilities. For example, the Tamworth may outcompete industrial swine stocks under pasture management, and breed-specific research would be a useful tool in determining the utility of various breeds in extensive production.

Mulefoot hogs are in critical need of conservation.

On pasture or confined in pens, pigs require tight, sturdy fences, and electric fences are often used. Strong and curious, pigs can wreak havoc on poorly constructed or ill-planned facilities. Pens or pastures should always include shade, since pigs cannot sweat. They wallow in mud in order to stay cool. Pigs keep themselves clean and have defined areas for eating and elimination, so rectangular spaces are best. Odors can generally be controlled by giving the animals well designed housing, good management and enough room.

Swine reproduction is nonseasonal for the most part. Gestation is about 120 days. Females, or sows, usually produce two litters of seven to twelve young annually. Males, or boars, can be aggressive and require skilled management. Pigs have productive lives of four to six years, although with good care they can easily exceed this lifespan.

While the national market for pork is strong, local markets are highly variable. Vertical integration of the swine industry means that a few corporations have interests in all aspects of pork production, from breeding to processing to marketing, with farmers contracting to grow the company's hogs. Corporate swine production has flooded the market with inexpensive pork, substantially lowering prices, and thus the income, for independent swine farmers in several parts of the country.

Independent swine farmers generally sell pigs at auction, and the prices they receive may also be low when any one company is the primary purchaser of stock. Those pigs which deviate from the standard profile will also be given a lower price. Access to processing plants is very important. If independent farmers do not have access to processing, they cannot sell high value, specialty pork directly to consumers who will pay for it. Without a strong market and consumer demand for extensively and independently produced pork, rare breeds of pigs are not likely to survive.

Poultry: Chickens, Ducks, Geese, and Turkeys

The domestic poultry species in North America include chickens, ducks, geese, and turkeys. The relatively small size of these birds means they require less space, cost, and labor than do the four-legged animals, yet there are many breeds in urgent need of conservation. Because of the importance of poultry, and of the many similarities in husbandry of these species, some general comments are made before each species is discussed.

Poultry, like swine, have always been an essential part of the diversified farm. Birds readily convert household, dairy, garden, and orchard waste into meat and eggs, and they are useful in the control of insect pests. The use of poultry in pastures with cattle and other large animals will greatly reduce the parasite and fly load as the birds eat insects and break up the manure piles that harbor larvae. Birds can also be used to control weed and insect pests in gardens, as well as to process compost and prepare beds for planting. Geese have historically been used as weeders for such plants as cotton and berries, as well as in nut groves and fruit orchards.

All poultry species are vulnerable to predators: domestic and wild dogs, foxes, owls, raccoons, weasels, snakes and others. Poultry conservation projects should be carefully protected from predators with secure overnight housing, fencing, and other measures. Even so, genetically valuable stocks should still be split into two or more flocks in different locations. A network of breeders serves the same purpose.

Poultry conservation projects can be fun for children. This girl is holding a New Hampshire rooster.

Poultry husbandry is not difficult, and there are many books available on the topic. Poultry conservation is, however, challenging. The survival of a large number of poultry breeds depends upon the abilities of breeders who are able to maintain a relatively large population size and hatch out replacement offspring each generation. It is ironic that these smallest of livestock require the most technically complicated conservation programs!

Poultry genetic conservation in the U.S. is focused on the standardized breeds, which include both production and exhibition types. (Landrace and feral stocks are not significant in North America.) Within the standardized breeds, however, there are sometimes confusing distinctions made. The

Dominique chickens are active foragers.

production breeds of poultry, sometimes referred to as "classic" breeds, have been traditionally selected for production of eggs, meat, or both. These breeds, particularly chickens, were the major producers of eggs and meat across North America as recently as the 1940s. They are true breeds in the genetic sense and thus are a high conservation priority.

In contrast, standardized breeds used for exhibition have been selected for perfection of form relative to the breed standard. Fanciers seek to breed individual birds which will win in the show ring, and crossbreeding (as well as linecrossing) may be used to produce these individuals or to renew vigor in exhibition stocks. As a result, exhibition poultry are not always considered breeds in the genetic sense of the word. Nevertheless, the continued activity of poultry fanciers is an important endeavor in the overall conservation of genetic diversity within the poultry species. For example, the Cornish, considered "only" a fancy breed for many years, is used today in the production of almost all broiler chickens.

Chickens

A wide array of chicken breeds is found in America, and most of these are endangered to some degree. The breeds vary according to size, conformation, color, plumage type, comb, and other physical features, as well as production capability and disposition.

Chickens are used to produce eggs and meat. Home grown chicken eggs show brighter orange yolk color and have richer flavor than supermarket eggs. They are easy to market and may be sold directly to consumers (including friends and neighbors) and restaurants. The sale of eggs is an easy accompaniment to the sale of vegetables, herbs, and other products, giving chickens excellent potential in market gardening businesses.

All breeds of chickens produce eggs, though if eggs are the goal of the enterprise, either dual-purpose or egg laying breeds are the best choice. Leghorns, Anconas, Minorcas, and other breeds of Mediterranean origin are excellent producers of white eggs, with commercial stocks of White Leghorns laying up to 300 eggs per year. These

Fresh eggs are one of the rewards of keeping chickens.

breeds also demonstrate a higher feed efficiency than brown egg layers. They are very active foragers, but fly so well that free range management can be a challenge.

Dual-purpose breeds, such as the Barred Plymouth Rock, Rhode Island Red, and White Wyandotte produce 250-280 brown eggs per year. These birds eat more feed and produce fewer eggs than do the Mediterranean breeds, but their larger frame makes them more useful for meat. They also tend to be quieter in disposition than the Mediterranean breeds. Dual-purpose breeds are still used for the production of commercial brown eggs, and they are often seen in free range egg production enterprises.

The status of meat breeds is, however, quite different. Several obstacles discourage the use of rare breeds in the production of meat. Industry-produced chicken is inexpensive and universally available. Though there is a growing demand for high quality, "free range" chicken, independent poultry producers in many parts of the country lack access to processing facilities. Direct sales to consumers are thus limited to the number of birds that can be processed at home. In addition, most states strictly regulate the sale of home-processed meat products and may limit the total amount sold.

A single type, the Cornish-White Rock cross, completely dominates poultry meat production in America. This bird grows very quickly to market weight and has soft, tender meat. As an intensively selected hybrid, the Cornish cross is not adapted to extensive husbandry; it is still the choice of small, independent producers who believe it to be the only acceptable chicken for the market. These producers nurse Cornish-cross stocks under extensive production, even though these birds lack the basic health, hardiness, and soundness of other breeds. Few producers ever consider other breeds, though these may provide better health and adaptability. The Delaware, Jersey Giant, and other breeds historically used for the production of meat prior to the Cornish-Rock are now in dire straits. Without the use of rare breeds in extensive enterprises, the Cornish-Rock will soon be the only meat bird left in North America.

Chickens mature rapidly and have a shorter lifespan than other livestock. Hens generally begin to lay at about five months old (depending upon day length), and they will lay eggs with or without roosters. Roosters are, of course, necessary for the production of fertile eggs and offspring. Chicken eggs take twenty-one days to hatch and chicks are brooded for four to six weeks until feathered out, when they may be called "fledglings."

Hens produce the most eggs during their first year, and they are only reasonably productive their second. After that, egg production falls rapidly. Chickens used for meat are usually butchered at twelve to twenty weeks, depending upon the breed and the owner's preference. (In contrast, Cornish-Rock crosses are ready for market at seven to eight weeks.) The natural lifespan of chickens may be four years, though most birds do not survive that long. Roosters offer some protection for the hens when the birds are kept out of doors, though their crowing can disturb neighbors.

Ducks

Ducks are among the most versatile and useful of all domestic animals. Ducks (and geese) are classified as waterfowl, and they do love water, but swimming ponds are not necessary for them to be raised successfully.

The Khaki Campbell is an excellent egg producer and forager.

Resistant to disease, tolerant of both cold and hot climates, and thriving wherever it is wet, ducks are easy to raise. They are very efficient producers of animal protein in the form of eggs and meat. Breeds selected for egg production can lay at a comparable rate (275-325 eggs per year) to the best laying chicken breeds. The eggs are excellent for baking and eating, and duck eggs may bring a good return when sold to ethnic markets or specialty bakers. There are limited markets for duck meat, but these may be developed according to the availability of processing facilities.

Ducks make good foragers, and the egg-producing and bantam breeds are the most active. David Holderread, who has written extensively about the utility of waterfowl, notes that ducks eat flies, Japanese beetles, grasshoppers, mosquitos, slugs, and snails (see *Raising the Home Duck Flock* by Dave Holderread, 1978, Storey/Gardenway). Ducks will also eat unwanted water plants (such as green algae, duckweed, pondweed, muskgrass, and wild celery) in ponds, lakes, and streams, and thus improve conditions for many types of fish. Holderread estimates that two to six ducks per acre of land, or six to ten ducks per acre of water will control these pests. Along with insects, plants and household food waste, a small amount of grain will give ducks a balanced diet.

Ducks are small in size and rarely aggressive toward humans. Their enthusiasm for life is quite entertaining, and the wide array of breeds presents many attractive choices. Ducks are productive for three to four years and can live up to six to eight years. Eggs take twenty-eight days to hatch, and they may be hatched by ducks or hens, or in an incubator. Like chicks, ducklings must be brooded until feathered. Adult ducks need shelter from wind and rain, as well as protection from predators.

Geese

Geese are hardy, economical, and useful farm birds. Like ducks, geese are easy to raise. They are relatively resistant to diseases and parasites, tolerate both cold and hot climates and prosper under wet conditions. These birds are the only true grazers of the poultry species and can thrive on a diet of grass, grit, and water. Fattening geese, however, may require feeding grain for four or five months per year in colder climates. Geese can easily be managed with ducks, and they make many a farm pond more beautiful.

The Pilgrim is an American breed that is now rare.

Geese are large and strong, with some breeds weighing twenty or more pounds. Geese can be aggressive and territorial, especially when defending a nest. An understanding of goose behavior will avoid most problems, however, and some breeds (such as the Pilgrim) are particularly known for having a calm disposition. All geese are alert, and they are sometimes used as "watchdogs" for this reason.

Of great interest on diversified farms is the utility of geese in pest and grass control. Geese keep wet, marshy lands from becoming overgrown, and they can be turned out in fields after harvest to glean grain seeds and eat pest plants. They prefer immature grasses and have traditionally been used as weeders in row crops and in orchards. When managed properly, geese are also excellent lawnmowers, useful for control of grass on pond and stream banks and of some underwater aquatic plants, such as pondweed, widgeon grass, bulrush, and eelgrass.

Geese are used to supply meat, eggs, feathers, and down. The eggs are considered excellent for baking and goose meat is traditional holiday fare. Feathers and down are valuable products of both geese and ducks. They can be harvested at the time of slaughter or annually from live birds during molting. Geese are long-lived, fifteen to twenty years, although peak production comes between three to five years old. Goose eggs may be hatched under a goose or in an incubator. Offspring hatch in about thirty days and are brooded until feathered.

Turkeys

Turkeys are large birds, and they require plenty of space. As their feed efficiency is lower than that of chickens, pasture production is an economical and healthy husbandry method for turkeys. Until the middle of this century, all turkeys were raised out of doors, often in rotation with crops or cattle. They have some utility in the control of insect pests. Industrialization, however, has transformed both the production method and the bird. Almost all turkeys alive today are of a single variety—the Broad Breasted White. All other varieties are rare. This poses a serious risk to the genetic health of the species, and turkey conservation is urgently needed.

The bronze turkey is named for the metallic sheen of its feathers.

The turkey is classified by the American Poultry Association as a single breed, with eight varieties. (These varieties are breeds in the genetic sense.) Turkeys range from the small sized Royal Palm, weighing twelve to twenty pounds, to the improved Bronze and White Holland breeds, with toms weighing thirty-five pounds or more. Turkeys show an astonishing range of hues, harkening back to the selective breeding for color practiced by the Aztecs centuries ago. Varieties also show differences in disease resistance, hardiness, and even intelligence, with many of the rare, less "improved" birds far better equipped for extensive husbandry. Some bloodlines, such as the Wishard line of Bronze turkeys, have been specifically selected to thrive under range conditions, and these are a high conservation priority.

Turkey meat can bring a higher return than does chicken, and the demand for high-quality turkey is increasing. The resources necessary for turkey management, however, are greater than that for the other poultry species.

Lack of access to processing facilities often limits the production of turkeys for direct and specialty markets. Where processing is available, the use of hardier rare varieties should increase the success of free range turkey enterprises and at the same time provide habitat for these endangered birds.

Domestic turkeys, including domestic bred wild turkeys, are long divergent from true wild turkeys. Domestic birds should not be released into the wild as interbreeding will compromise the genetic integrity of the wild population.

In general, turkeys require a higher level of husbandry and hygiene than do chickens, and mixing of the species is not generally recommended. Turkeys are most productive in their first year of laying, with production declining rapidly through the second and third years to become negligible. General vigor of the birds also declines, and most are quite elderly by five years old. Eggs are incubated for twenty-eight days and birds are brooded about eight weeks until feathered.

Chapter 6. Choosing a Breed

Dominique chicken

The choice of a specific breed is a highly individual decision.The right choice is essential to the success of a breeding program. Conservation breeding is a long term commitment, and it is vital that the breeder enjoy the animals. Certain people and certain breeds seem to "click," and that is the goal when selecting a breed. Conservation breeding is important work, but it should also be fun!

When choosing a breed, one must consider breed characteristics, uses, husbandry requirements, farm resources, and regional and climatic adaptations. Information about breeds is available from a variety of sources, including publications, breed associations, and, most importantly, other people. Prospective breeders should visit several herds or flocks before choosing a breed and selecting stock.

All breeds of a species require the same general husbandry, though the inputs and management required may vary. For example, all horses need the same minimum facilities and feeds, though the amounts and mix will be different according to the breed. Some breeds are highly selected for specialized tasks; new breeders are needed to continue finely tuned selection programs for a specific array of traits. Other breeds are primitive, having been shaped by natural selection, and they need a more "hands off" approach. Most breeds fall between these extremes.

Aspects of veterinary care may vary somewhat for the different breeds, but not as widely as is commonly assumed. The establishment of a health plan in consultation with a veterinarian is necessary for every herd or flock. Preventive veterinary care, such as vaccinations and dewormings, is recommended and almost always necessary.

The Jacob sheep has attracted a dedicated and skilled group of breeders

To ignore this is to court disaster. While some breeds demonstrate parasite and disease resistance in certain conditions, all can succumb if not intelligently managed. Veterinary care is especially necessary for the rarest of the rare. In such conservation programs, it may be essential not to lose any individual animals. Selection for disease and parasite resistance can always come later when numbers have been increased.

Regional and climatic adaptations are important to consider. Highland cattle, for example, are adapted to cold climates, and this breed would be a poor choice for the heat and humidity of the deep South. It is advisable to look for breeds which are appropriate for one's climate and to the production system planned. A network of breeders may already be in place, facilitating breeding and marketing of stock.

An obvious consideration in breed selection is breed characteristics, that is, size, shape, disposition, and uses. As with the choice of species, a breeder needs to determine what role the animals are to play in a larger agricultural plan. Some breeders may want to consider the agricultural history of their area and choose breeds which have been documented as appropriate in that place. These breeds may also have added market value because of their historical status. Dominique chickens, for example, generate interest in the southeastern U.S. far beyond their production capability, simply because these are the chickens that nearly everyone's grandmother raised.

Each breed can be used to produce some product, even if it is simply offspring. Some breeds produce specialty products, such as Jacob fleeces or Highland beef, but marketing research and a business plan are essential to generate income. For people without the time or interest to develop such markets it may be preferable to select a breed whose products fit the existing marketing structure for more generic livestock or poultry.

Finally, breeders should consider which type among the various sizes, styles, and specializations *within* a breed is most appropriate and interesting. Differences within breeds have been created by selection, sometimes over a long period of time; the results are often described as traditional versus modern type. Purebred breeding occurs in a marketplace of ideas, and ideas can transform a breed slightly or dramatically.

It is often the case that traditional types, those least "modern" or "improved," are most in need of foresighted genetic conservation programs. Fashions in all breeds will go through cycles. Traditionalist breeders who can survive through thin times will usually have others come to appreciate the historic types of animals once again. This has been demonstrated in cattle breeding, where large framed cattle were penalized in the fifties and sixties only to regain popularity in the eighties and nineties. The Suffolk, a draft horse of traditional farm type, is now regaining popularity as breeders recognize its utility compared to the parade and show types increasingly common among the other draft horse breeds.

Highland cattle thrive in cold climates.

The choice of a breed does not necessarily require that all of these decisions be made at once. In fact, most breeders are likely to change the focus of their conservation work within a breed as their expertise increases, and this should be encouraged.

American Breeds in Need of Conservation

The American Livestock Breeds Conservancy does a periodic inventory of breeds in North America to determine which are in need of conservation. A summary of the most recent North American livestock census appears in Figure 6.1 and the most recent poultry census in Figure 6.2.

Breeds which appear on these conservation priority lists are rare in the U.S. and globally. Of greatest concern for ALBC are those breeds which are unique to America and those international breeds for which American breeders can make a difference.

Conservation programs for endangered breeds are described in Section III. Critically rare breeds are in urgent need of a rescue program to increase numbers and decrease relatedness among individuals of the breed. As numbers increase, the rescue protocol can be replaced with a conservation program to expand the population further and maintain the diversity established during the rescue phase.

A less urgent situation exists for rare breeds where numbers are higher and where multiple bloodlines exist. Conservation breeding is easiest in this situation, and there is also the luxury of some selection for performance. At the same time, continued use of multiple, unrelated males is necessary at both the breed and herd or flock level to avoid any loss of variability and consequent viability. In this situation, the work of several breeders can have dramatic results in the increase of breed numbers and conservation of genetic breadth.

Conservation programs for the most numerous of the rare breeds may focus on specific bloodlines or types within the breed, and such programs frequently include improvement. Such programs are significant where there has been genetic erosion within the breed.

Breeders should not choose a conservation project based solely on its urgency, nor should they compare their project to others. Each breed on the ALBC conservation priority list is worth an investment of time and resources. There is plenty of work to do if the long term survival of all rare breeds is to be assured.

Figure 6.1 The American Livestock Breeds Conservancy priority list of livestock breeds

This list resulted from a census of U.S. and Canadian livestock breeds conducted by ALBC in 1992-1994. It was first published as part of *Taking Stock: The North American Livestock Census* by D.E. Bixby, C.J. Christman, C.J. Ehrman, and D.P. Sponenberg (1994, McDonald and Woodward Publishing Company, Blacksburg, VA). The list was updated in 1995.

Critical: Breeds with fewer than 200 annual North American registrations and an estimated global population of fewer than 2,000.

Rare: Breeds with fewer than 1,000 annual North American registrations and an estimated global population of fewer than 5,000.

Watch: Breeds with fewer than 2,500 annual North American registrations and an estimated global population of fewer than 10,000. Also included are dairy cattle breeds which have shown a marked decline in numbers over the last twenty years.

Study: Breeds which are of genetic interest but lack documentation or definition.

	Critical	Rare	Watch	Study
Asses	American Mammoth Jackstock* Poitou		Miniature Donkey	
Cattle	Ankole-Watusi Devon Dutch Belted Florida Cracker* Kerry Milking Devon* Pineywoods* Randall Lineback* White Park	Belted Galloway Dexter Galloway	Ayrshire British White Guernsey Highland Milking Shorthorn Red Poll	
Goats	San Clemente (F)*	Myotonic* (Tennessee Fainting/ Wooden Leg) Nigerian Dwarf	Oberhasli Spanish*	
Horses	Akhal-Teke American Cream* Caspian Cleveland Bay Exmoor Mountain Pleasure* Rocky Mountain* Suffolk Punch	Clydesdale Dartmoor Lipizzan Spanish Mustang/ Spanish Barb*	Friesian Gotland Hackney Horse Shire	
Sheep	Gulf Coast Native* Hog Island (F)* Leicester Longwool Santa Cruz (F)* Wiltshire Horn	Cotswold Delaine Merino Karakul Katahdin* Navajo-Churro* St. Croix* Tunis*	Black Welsh Mtn Clun Forest Dorset Horn Jacob Oxford Shetland	Caribbean Hair
Swine	Gloucester Old Spot Large Black Mulefoot* Ossabaw Island (F)* Saddleback	Hereford*	Tamworth	Choctaw* Guinea Hog* Red Wattle*

*Unique to North America

(F) Feral population or breed of feral origin

Figure 6.2 American Livestock Breeds Conservancy priority list of poultry breeds

This list was generated by a survey of U.S. poultry breeds conducted by the American Livestock Breeds Conservancy in 1987. Survey results were first published in 1988 by the American Livestock Breeds Conservancy in the *AMBC Poultry Census and Sourcebook* by L. Heise and C. Christman. The list includes breeds which have been of cultural, historical, or economic importance in America.

Critical: Breeds with fewer than 500 breeding females and fewer than three primary sources.

Rare: Breeds with fewer than 2,000 breeding females and five or fewer sources, or a concentration of the breeding population in three sources.

Watch: Breeds with fewer than 10,000 breeding females, but fewer than ten sources or a declining number of sources.

Species	**Critical**	**Rare**	**Watch**
Chicken	Ancona Black Minorca Delaware Dominique White Jersey Giant White Wyandotte	Black Jersey Giant Brown Leghorn	Barred Plymouth Rock Black Australorp New Hampshire Rhode Island Red
Duck		Khaki Campbell	Rouen
Goose	Pilgrim		Toulouse
Turkey	Bronze (unimproved)		

Acquisition of Stock

After selecting a breed, it is time to begin locating individual animals. Details become important: what age? what sex? how many of each? The answer comes from the immediate uses for the animals and the scope and pace of the program undertaken.

Prospective breeders should shop around for stock. Acquisition of animals is an important investment, so it is advisable to talk to several reputable breeders and look for animals appropriate to the management system planned.

It is desirable that the stock be matched to the experience level of the breeder. Individual animals critical for genetic conservation should generally be in the hands of experienced rather than novice breeders. An inexperienced person may find it worthwhile to purchase well started or adult animals which will require less specialized husbandry than young stock. An exception is poultry, where even the inexperienced breeder will usually begin with day-old birds.

Another decision to be made is whether to buy registered or unregistered livestock. Animals are "registered" in breed registries. These organizations maintain records on all purebred individuals within a breed. Registration is recognition that an given individual is part of a breed, and it is given a certificate, often referred to as "papers." An animal's registration certificate includes its pedigree (parentage) as well as relevant descriptive information, including any permanent ID, such as a brand, ear tag, ear notch, or tattoo.

Registered stock, such as these Tamworth piglets, is necessary for conservation programs.

It is almost always the case that investment in registered stock is necessary to develop a conservation breeding program. Poultry are not registered, but the purchaser should seek documented, purebred birds from a reputable breeder.

Prospective buyers should be sure to find out the registration status of an animal and to match up the permanent identification on the registration papers with the animal presented. Upon sale, it is the responsibility of the *seller* to "transfer" ownership of the animal to the buyer by sending the necessary paperwork and fees to the breed registry. In the excitement of acquiring a new animal, it can be easy for buyers to overlook the paperwork associated with a transfer, and sellers are not always prompt about getting this work done once the animal is sold. Buyers should insist on a formal bill of sale and the prompt transfer of papers, and they should never assume that a registry problem can be fixed later. If the transfer is not done properly, the new owner can find it difficult or even impossible to recover an animal's documentation.

Buyers should insist on a formal bill of sale and the prompt transfer of registration papers.

It is sometimes the case that a seller offers to sell a registered animal for a lower price if sold without its registration papers. This eliminates the individual animal and *all* of its offspring from the documented purebred population. In addition, lack of papers significantly lowers the value of the animal (as well as all of its offspring) at the time of next sale. This practice has no place in conservation breeding.

Most registries have closed books, which means that *only* offspring of registered parents can be registered. This is important to note. While purebred, unregistered individuals can be genetically interesting or even important, if they cannot be registered, such animals are of no practical value to a purebred breeding program.

Other breed registries are open, which means that animals can be registered based on inspection and other documentation whether or not their parents are registered. A buyer should carefully check the rules to be sure that the individual animal can be registered before a commitment is made to purchase.

Interstate shipment of animals requires appropriate health testing and a health certificate to signify disease-free status. These requirements are different for each species and in each state. Veterinarians can provide information about shipping requirements since they must prepare and sign health certificates. It makes sense for the seller to supply a health certificate even for in-state sales, and some states require that health papers be presented at time of sale. For example, proof of a negative Coggins test (for Equine Infectious Anemia) may be required for sale of a horse within some states as well as to cross state lines.

Shire horse

Chapter 7. Breeder Associations and Networks

A person who becomes a livestock breeder also becomes a part of a community of livestock breeders. This community usually includes a breed association (or society), which is an organization of individuals who own, admire, and promote the breed. The association facilitates communication between breeders and allows individuals to participate in making policy for the breed. All breeders of a given breed should be members of its association. Active participation is necessary if one is to encourage others to become conservation breeders.

In some instances two or more associations exist for a single breed. Split of a breeder community is often caused by a dispute over breed registry rules. While this cannot always be avoided, a split is generally to the detriment of the breed as it disperses breeders' resources and may lead to a split in the breed's gene pool.

A major responsibility of a breed association is the operation of a registry of animals within the breed, with rules for their registration. Registries maintain pedigree information on all animals and as a result, have records for the breed as a whole. It is important that breeders register their animals; failure to do this is a disservice to the breed. Nonregistered, undocumented animals are wasted, lost both to a breed's history and to its future. Some breeders choose to register only those animals used for breeding and skip the registration of stock to be culled. This is not necessarily a problem as long as complete farm records are kept for all animals.

Though the breed association is the formal tie between individual breeders, regional associations, clubs, and informal networks of breeders are also important

Breeder networks, such as those for Dutch Belted and Dexter cattle, are essential for breed survival.

networks of communication. Such networks may serve to keep people informed of developments and trends within the breed and facilitate exchange of stock. This exchange is often the basis of long-lasting friendships as well. It takes time and consistent participation for a newcomer to become a part of an informal network, but this is because it takes time for a new breeder to become aware of the people involved, their histories with the breed, and their attitudes and actions. Long term commitment to a breed and to the breeder community is quite different from the more transitory activity typical of the speculator. Speculators damage a breed by buying up good genetic material only to discard or disperse it haphazardly once the initial thrill wanes. Such activity is especially common in breeds for which there is a ready market and the appearance of quick money to be made.

Individual breeders and breeds are also served by the breeder community at the time of dispersal of a herd or flock. This can happen as a result of an emergency or disaster, or at the end of a breeder's career. In the latter, the final dispersal after a long, successful breeding program becomes an important event for the breed. A strong breed network can keep the stock in the hands of other dedicated breeders, so that the genetic wealth it represents is passed along to those who will appreciate it. In contrast, if animals are dispersed only through commercial channels, they are usually lost to the breed, and this can have dire consequences.

To protect the future of their own herds or flocks, breeders should make provisions for their stock in written instructions given to a family member or friend. These instructions should include specific plans for all animals and resources set aside for their accomplishment. Care for stock can also be put into a will, though animals will likely be dispersed long before a will is probated. Foresight can make a tremendous difference in allowing one's colleagues to carry on the work of breeding and conservation.

Section III.

Conservation Programs

Rouen ducks

Chapter 8. The Context of Conservation Breeding

The American Livestock Breeds Conservancy encourages the protection of genetic diversity through the effective conservation of livestock and poultry breeds. The diversity represented by many genetically distinct breeds is necessary for the long term health and utility of each domestic species. Diversity within each breed is necessary for the breed's long-term well-being and commercial potential.

The ALBC has designed conservation breeding programs for livestock and poultry that rely on the science of genetics as well as practical management considerations. Livestock and poultry breed conservation is firmly based in science, but practical considerations rapidly make or break a program. The following programs are designed for modification to fit the various species and breeds. Use of the programs will strengthen the skills of individual breed stewards.

Chapter 8 includes a discussion of the context for conservation, including the measures of breed status. Chapter 9 discuses the basic conservation program for breeds of livestock which are not critically rare. A rescue program for critically rare populations is described in Chapter 10. Poultry conservation breeding is considered in Chapter 11.

Breed conservation has five elements, each one essential for success:

1. A knowledge of breed history and characteristics, including production parameters.
2. An evaluation of the numerical and genetic status of the breed to determine

This photograph of a Florida Cracker cow was taken in 1926. The breed looks much the same today.

its needs, both in terms of numbers and maintenance of existing genetic diversity. Distinct bloodlines or historic breed types that represent significant genetic resources within the breed should be documented and conserved.

3. A determination of the place of the individual herd or flock within the breed, with an understanding of how the specific herd or flock best serves the long term genetic needs and existing network of the breed.
4. The design of a breeding program for a specific herd or flock, using the ALBC conservation programs that follow. These emphasize:
   - the use of several sires to maintain the genetic variation of the population, as represented by its distinct bloodlines;
   - the management of inbreeding so that it does not affect the entire herd or flock;
   - the selection for breed type, soundness, production, and also for improvement when appropriate.
5. A periodic reevaluation of the numerical and genetic status of the breed, as well as the herd or flock, and revision of the breeding program as needed.

These are the basic principles illustrated in the programs to follow. There is no single, simple recipe for conservation breeding which can be applied to every herd and flock of all species and guarantee that genetic diversity is protected. Nor is such a recipe necessary for breeds to be conserved. The goal is not uniformity, either of breeds or breeders. If breeders understand the principles and methods of conservation breeding described, they will be able to develop programs to fit their own breeds and situations. A diversity of well-considered approaches within the conservation philosophy is healthy for both a breed and its stewards.

Measures of Breed Status

Several methods have been developed to describe the numerical and genetic status of animal populations, and these can be applied to breeds. The utility of these methods varies according to the types of breeds surveyed and the accuracy of the information available about them.

The Florida Cracker breed includes several distinct bloodlines, among them the Barnes line which is conserved by Calvin Hutchins.

Numerical status

The numerical status of a breed can be considered first, that is, how many individuals of the breed exist? One can answer this question by estimating the number of adults or the number of adult males and females. Such information, however, is usually difficult to come by. Since breed registries do not record deaths of individuals, the registry may have a cumulative total of all individuals which have been registered but not know the number still alive. It is therefore impossible to verify population estimates, to compare them with each other, or to describe trends within a breed. In addition, a population estimate by itself gives no indication of the number of males and females that are being used for purebred breeding and therefore making a contribution to the next generation.

A better measure of the numerical status of breeds is to consider the number of purebred offspring produced each year. The American Livestock Breeds Conservancy uses the number of purebred offspring registered annually as the best estimate of the numerical status of a breed. This measures only those individuals which are likely to be used for purebred breeding and thus begins to estimate the size of the next generation. This figure also provides a close approximation of the number of females used for purebred breeding each year. An additional advantage is the accuracy of registration figures, which are available from breed registries and can be compared year to year to monitor breed trends.

Numerical and genetic status: effective population size

When the exact number of male and female parents used to produce a given generation of a breed is known, then a numerical calculation called "effective population size" can be obtained.

A simplified equation to describe effective population size is:

$$\mathbf{Ne} = \frac{[4 \times \mathbf{Nm} \times \mathbf{Nf}]}{[\mathbf{Nm} + \mathbf{Nf}]}$$

Ne = the effective population size; **Nm** = number of male animals; and **Nf** = number of female animals.

The effective population size equation demonstrates that the size of a population does not necessarily reflect its genetic status, and an increase in numbers alone will not necessarily protect genetic diversity. Instead, the number of parents used to produce each generation is critical. The number of male parents is most relevant, since generally fewer males than females are used in breeding. If only a few males are used to produce a generation of offspring, genetic breadth will be substantially reduced, regardless of the number of females used.

The size of a population does not necessarily reflect its genetic status.

The use of a single male parent dramatically reduces effective population size, because the generation produced by a single male is composed of half-siblings with limited genetic variability. Figure 8.1 makes obvious the consequences of using a single sire, especially if he is used for several years in a row, then replaced by his son, and so forth.

A disadvantage in the use of effective population size is that its calculation becomes much more complicated under "real world" conditions. If generations overlap, if generations are mated to one another, if mating groups differ in size, or if some individuals reproduce less than others—all conditions usually true of livestock and frequently true of poultry—then the calculation of effective population size rapidly becomes difficult. In addition, effective population size makes no correction for information about the relatedness of the male parents. It is quite different (and not terribly useful) to use many males which are closely related than to use the same number of unrelated males.

Inbreeding coefficient

Changes in the genetic status of a population, particularly the differences between effective population sizes, are often measured using the inbreeding coefficient. The inbreeding coefficient is a geneticist's calculation of the proportion of genetic variants (or alleles) in an individual that are likely to be identical. They are identical because the sire and dam are related. The more closely the sire and dam are related to each other, the larger the number of identical alleles that are passed onto the offspring. The inbreeding coefficient is therefore a measure of the level of genetic uniformity that comes directly from inbreeding.

The equation for inbreeding coefficient of direct concern for rare breeds is:

$$\textbf{The change in inbreeding per generation} = \frac{1}{8 \times Nm} + \frac{1}{8 \times Nf}$$

Several examples of inbreeding coefficient appear in Figure 8.1. These calculations are made based on the assumption that every male produces a son for the next generation and every female a daughter. If this does not occur, then the increase in inbreeding will be higher than that listed in the table.

Figure 8.1 Effect of sex ratio and population size on inbreeding. Actual population size, effective population size, and changes in inbreeding coefficients per generation are given to illustrate the results of using different numbers of male and female parents to produce a generation of animals.

Number of males	Number of females	Total population	Effective population size	Increase in the inbreeding coefficient per generation
1	30	31	3.87	12.9%
3	30	33	10.9	4.6%
9	30	39	27.6	1.8%
30	30	60	60	0.82%
1	60	61	3.934	12.7%
1	90	91	3.96	12.6%

The usual recommendation for long term maintenance of populations is that increases in inbreeding be kept below 5% per generation. In small populations the rate of increase can easily go much higher than this, and the result can be a decline in reproductive fitness and general vigor. An increase in inbreeding during one generation cannot be corrected by increasing numbers of animals in the next generation. The only way to decrease the inbreeding coefficient in a group of animals is to bring in new animals.

The number of parents of the less numerous sex (usually males) is the limiting factor in effective population size and also the main contributor to higher rates of inbreeding. Increasing the number of male parents, even modestly, has more genetic impact than a relatively great increase in female numbers.

Like effective population size, the inbreeding coefficient is primarily a theoretical measure. Its primary use is to compare different breeding options at the herd or flock level. Since breeds vary in their tolerance of inbreeding and past selection against inbreeding depression, it is not necessarily useful to compare inbreeding coefficients between breeds.

Heritage, habitat, and other factors

Despite the utility of mathematical calculations in demonstrating genetic principles, the genetic status of a living breed is very difficult to describe and will always remain somewhat subjective. Each breed has its distinct genetic heritage and its own habitat, and these must be considered when describing its genetic status.

Each breed has its distinct genetic heritage and its own habitat, and these must be considered when describing its genetic status.

Among the elements of a breed's genetic heritage is the existence (or lack thereof) of distinct, isolated bloodlines, which have high genetic value for a breed. These lines represent genetic material which has been long divergent from the breed as a whole and can be used to provide linecrossing and some hybrid vigor. Also of interest is the recent history of the breed and its relative resistance to inbreeding.

Included in any measure of breed status is the status of its habitat. Like wild species, domestic animals are dependent on habitat for survival. Unlike wild species, the habitat for livestock and poultry breeds is made by both nature and human breeders. One element is geographical distribution. If all individuals of a breed are found in a single location, then disease, predation, or environmental disaster can have a devastating impact on the breed. In contrast, a breed that is dispersed throughout several locations it is less vulnerable to loss. The number and location of herds or flocks is an especially important indicator of status for those stocks most vulnerable to predation—poultry and small ruminants—but it is an issue for all species.

Milking Devon cattle have been protected from extinction by New England teamsters and dairy farmers.

A second element of habitat is the selection of breeding stock practiced by human breeders, and the ways that selection is evolving to fit market conditions. What are breeders' goals for the breed? It is important to know whether current selection pressures fit the historical background of a breed. Intense selection that seeks to mold the breed into a different direction can cause genetic erosion. Breeds which have a current market niche consistent with their historical uses are in a far better position than those which lack this opportunity. Distribution is also relevant. If a breed population is divided among only a few herds or flocks, then it can readily suffer genetic erosion due to poorly designed selection programs.

The network of people who promote a breed, as well as the resources available to the breed, are an important part of its habitat. The strength or weakness of a breed association and the vision of a breeder community serve to make a breed either more or less secure.

Describing the status of a breed, therefore, is a complex task. A breed may appear numerically strong, but the use of only a few sires each generation has made it genetically fragile. A breed may have a relatively large population, but may be in danger as its purebred females are being siphoned off into crossbreeding programs. A breed may be rare, but have several distinct bloodlines in the hands of dedicated breeders, giving it a stronger base than the numbers alone would dictate. This is quite different from the situation facing a rare breed which is limited to one location and selection program. The complexity of breed status—and the fact that it can change quickly—is another reason that all rare breeds are in need of conservation programs, and that no conservation work is wasted.

Tamworth pig

Chapter 9. The ALBC Conservation Breeding Program for Livestock

The ALBC conservation program described in this chapter can be applied to almost all of the livestock species. The program uses cattle as the example, and it is an easy fit for sheep, goats and swine. Horses and donkeys fit, though somewhat awkwardly. Conservation of poultry breeds, which has the same goals but requires different management strategies, is described in Chapter 11.

The livestock conservation program is explained in the text and illustrated in Figure 9.1. The objectives of the program can be summarized as the identification of distinct bloodlines in the herd and the maintenance of these bloodlines over several breeding cycles. This will be accomplished while managing the herd as a single unit.

The conservation breeding program begins with an evaluation of the relatedness of the herd population and the division of animals into three groups that are more closely related to each other than to the herd as a whole. Each group should be of approximately equal size, and each should have females and one or more males. The composition of the groups will vary depending upon the composition of the herd, and the method of grouping is not critical. Individuals may be divided according to their pedigrees, with all the descendants of a specific sire or dam put together in one group and descendants of another sire or dam in another group. Alternatively, if the herd has been built from several other foundation herds, then individuals can be grouped according to which foundation they represent. For practical purposes, these groups can be considered bloodlines, or lines, within the herd.

The Gloucester Old Spot is among the swine breeds in need of conservation.

The animals will be managed in such a way that all bloodlines are maintained, even though the animals are managed as a single herd. To accomplish this, males from the different lines will be rotated sequentially through the herd. The male from the first line will be used on the entire herd during the first mating season. In the second season, the male from the second line is used. The male from the third line is used in the third mating season.

Use of the three male lines sequentially allows the herd to be managed as a single unit, and most breeders will find this much easier than having multiple breeding groups in separate pastures. The different lines within the herd are separated by time (using the males sequentially) rather than space (splitting the herd and using males of all three lines on one third of the herd each cycle). If the herd is too large to be served by one male, then the management can shift to that of two parallel herds, a modification that is discussed later in this chapter.

The general term "cycle" is used to describe the consistent period of time for each cycle of reproduction, including breeding, gestation, and the birth of offspring, until the females are ready to breed again. In cattle, this cycle is approximately twelve months, with some variation according to breed. Many breeds of sheep and goats and nearly all swine will produce offspring more frequently than once a year, and thus their cycles are shorter. Horses and donkeys will produce offspring every year or longer, so that their cycles are somewhat longer.

Cycle 1: A herd of cattle is divided into three bloodlines of approximately the same size, labelled A, B, and C. Each line is made up of cows and at least one bull.

The bull from line A is bred to all of the cows (A,B, and C). The resulting calves are born about nine months later. Each calf is identified with an ear tag, brand, or other permanent marker. In addition, each calf is classified according to its parents' bloodlines. Since the father of all the calves is the A bull, and their mothers are either A, B, or C cows, the calves are, respectively, A/A (or simply A), A/B, or A/C. (For the purposes of this example, the sire's letter will be given first.)

All of the female calves (A, A/B, and A/C) will be kept. At maturity, which is a year

Figure 9.1 The ALBC conservation breeding program for livestock, using cattle as the example. The breeding cycle for cattle is approximately twelve months long. Offspring are produced at the end of the cycle, just before the next cycle begins. Note that female offspring are rotated into the breeding group at approximately one year old.

Cycle	Bull used	Cows bred	Offspring produced	
			Linebred, save males and females	Linecross, save only females
1	A	A, B, C	A	A/B, A/C
2	B	A, B, C	B	B/A, B/C
3	C	A, B, C, A/B, A/C	C (C/AC)	C/A, C/B, C/AB
4	A	A, B, C, A/B, A/C, B/A, B/C	A (A/AB, A/AC, A/BA)	A/B, A/C, A/BC
5	B	A, B, C, A/B, A/C B/A, B/C C/A, C/B, C/AB	B (B/AB, B/BA, B/BC, B/CB, B/CAB)	B/A, B/C, B/AC, B/CA
6	C	A, B, C, A/B, A/C, B/A, B/C, C/A, C/B, C/AB, A/BC	C (C/AC, C/BC, C/CA, C/CB, C/CAB, C/ABC)	C/A, C/B, C/AB, C/BA

to eighteen months old depending upon the breed, the breeder will select among the female calves and mate those which are sound and of breed type. The future A line bull will be selected from among the A/A bull calves. Though only a single bull will be

Breeders of critically rare White Park cattle are using a conservation plan for the breed.

needed, it is best to keep two or three A line bull calves as possible replacements and choose among them at maturity. This allows for selection and also insures against the loss of a bloodline sire if a single male dies from illness or accident. The A/B and A/C bull calves are not used for breeding within this herd, although might be of use in other herds. The mature A line bull, father of this crop of calves, is retired from the herd, having made his genetic contribution and having produced a replacement son.

Cycle 2: When the cows come back into heat, the B line bull is used on the entire herd. This is still early in the program, so the only females mated are the same mature A, B, and C cows used the cycle before.

Offspring produced nine months later represent B/B (or simply B), B/A and B/C. Female calves (B, B/A and B/C) are retained for possible breeding at maturity. The replacement for the B line bull will be chosen from among the B line bull calves, so one or more of these young males is retained. The B/A and B/C bull calves will not be used for mating within this herd. The mature B line bull is retired from the herd, having made his genetic contribution and having produced a replacement son.

Cycle 3: The herd now consists of: the mature A, B, and C cows; the mature C line bull; young cows A, A/B, and A/C (born in cycle 1) which should be mature and ready to be mated; heifer calves B, B/A and B/C just born and therefore not ready to be mated; and young bulls which are the next A line and B line sires. When the young A cows join the breeding group, they increase the number of cows in the A bloodline. Each line receives new females every third cycle.

The third breeding cycle begins when the cows come into heat after the cycle 2 calves are born. The C line bull is mated to all of the cows, that is A, B, C, A/B and A/C. The offspring born nine months later are: C/A, C/B, C, C/AB, and C/AC. This looks more confusing than it really is. Because the C/AC calves are more than one-half line C (they are actually three-quarters), they can be listed as C, simplifying the offspring list to the less complicated C/A, C/B, C, and C/AB.

As before, all female calves (C/A, C/B, C, and C/AB) are kept for possible breeding at maturity. The replacement for the C line bull is selected from among the C line bull calves, and the mature C bull is retired from the herd. Bull calves C/A and C/B are not retained.

Cycle 4: It is now time to again use a sire from line A. The A bull used is one of the young males saved from cycle one, now two years old. The cows in the herd are: A, B, C, A/B, and A/C. Young cows B/A and B/C, born in cycle 2, will be bred if they are mature. Offspring produced are: A, A/B, A/C, A/AB, A/AC, A/BA and A/BC. Again, classify those which are more than 1/2 line A as simply A, so the offspring group is reduced to A/B, A/C, A/BC and A (which includes A/AB, A/AC, A/BA).

Offspring produced each year are considered linebred (if more than 1/2 A, B, or C) or linecross (a mixture of A, B, and C with no one line over one-half). Both linebred and linecross animals are important to the success of the program. Linebred animals replenish the A, B, or C bloodlines. Linecross females will be used to produce replacement males or females for any of the lines, depending on which bull they are bred to (see Figure 9.3).

Alternation of inbreeding and outbreeding

The essential part of the conservation breeding program, from a biological standpoint, is that the genetic material in the herd is alternating between an inbred and an outbred condition. An example is given in Figure 9.2. Though this example does not track animals through consecutive mating cycles, it does illustrate the genetic contri-

Figure 9.2 Alternation of inbred and outbred conditions in a conservation herd of cattle

cow A x bull C	=	cow C/A (outbred)
cow C/A x bull A	=	bull A/CA (=A) (moderately inbred)
bull A x cow B	=	cow A/B (outbred)
the cow A/B x bull C	=	cow C/AB (outbred)
cow C/AB x bull A	=	cow A/CAB (=A) (minimally inbred)
cow A/CAB x bull A	=	bull A/ACAB (=A) (moderately inbred)

Careful planning gives each calf a role in conservation of the herd. These are Dexter cattle.

butions from each bloodline over several generations. The genetic material from the original A line cow swings back and forth from outbred (to the B and C lines) and to inbred (the A line) animals.

Moderate inbreeding is occurring in as many different directions as there are bloodlines (lines A, B, and C). The fact that inbreeding occurs in different directions means that the genetic distance between different lines is being maintained within the population. This distance provides the opportunity to relieve inbreeding if necessary. Animals that are inbred to the A line may be mated to a B line or a C line animal to produce an outbred offspring. In addition, the alternation of inbreeding and outbreeding in the conservation program regularly introduces hybrid vigor in the form of the outbred individuals.

Other breeding systems may seek to minimize inbreeding across an entire herd at once. Most of these, however, only assure that inbreeding is occurring slowly through the entire herd and always in the same direction. This leaves no outcross available within the herd.

In most rare breeds, some inbreeding will be unavoidable over the long term. This conservation breeding program constrains the population so that inbreeding is managed. It occurs in an orderly fashion and only in some parts of the population (the inbred A, B, and C animals). After many generations, some low to moderate inbreeding will exist across the herd, but it will be much lower than without the breeding program described.

Size of the herd

A herd of some size is necessary for the success of the ALBC conservation breeding program. There must be enough mature females in *each* of the bloodlines to produce a good replacement male and several replacement females *every breeding cycle*. Ten females per line is a minimum, and fewer than ten is probably playing the odds a little too close. The actual number required will vary according to how selective a breeder wants to be and how good a breeder's luck is. The three lines in our herd example, each with a minimum of ten cows, translates into a total herd of thirty cows.

Remember, however, that the outbred cows contribute replacement offspring to more than one bloodline. It is possible, therefore, to have only thirty cows represent three lines quite sufficiently. The cows that are linebred produce replacements only for their line, but the linecross cows can produce replacements for either of two lines. For example, linecross cow A/C can produce replacements for line A when bred to bull A and replacements for line C when bred to bull C. Each line will have contributions from fifteen females each, that is, five to seven females that are inbred to the line, and five to seven that are outbred from the line. Figure 9.3 provides an illustration.

Figure 9.3 Source of bloodline replacements in a conservation herd of cattle

Cows per line		**Source of replacements for each line**		
Line or linecross	**# of cows**	**A**	**B**	**C**
A	5	✓		
A/B	5	✓	✓	
A/C	5	✓		✓
B	5		✓	
B/C	5		✓	✓
C	5			✓
Total	30	15	15	15

In a given year, the cows are gestating calves of one line, nursing calves of another, and a bull of the third line is waiting in the wings to be used as a two year old. Bulls are used for mating when they are young, and then they are retired from the herd. Use of one bull on the whole herd makes it unnecessary to decide which cows to mate with which bulls. This is not just a lazy person's way out of work. The mating of specific individual males and females is usually done in anticipation or need of a male or a female offspring, and such outcomes are impossible to predict. Mating the entire herd sequentially to males of the different lines gets around this problem. With a large enough herd, there will always be enough males and females of each line produced in each breeding cycle to assure replacements and maintain each bloodline.

It is important to remember that the ALBC conservation breeding program is designed to protect genetic variation in a relatively small, contained population, whether that be a herd, a strain, or an entire breed. For larger populations (herds or breeds) in which more options for linecrossing and replacements are available, the constraints of the program can be relaxed. It may be less important that all sires be linebred, although it is still very important that multiple unrelated sires exist within the population so that the population does not become inbred in the same direction.

The ALBC conservation breeding program is designed to protect genetic variation in small, contained populations.

Because the program is designed for management of a self-contained population, it does require a herd or flock size of thirty or more females. Many breeders may not be able to keep herds or flocks of this size, but they should not be dissuaded from conservation. Breeders of smaller herds or flocks can form networks; through such management, numbers of animals may easily reach the totals described in the conservation program described above or in the program for parallel herds described below. Important progress can be made in small, carefully managed and mated herds and flocks, and this progress is important to the genetic status of a breed.

Beginning breeders will often find it desirable to build up herd or flock numbers gradually as their knowledge and experience increase. All breeders can apply general conservation principles regardless of the size of the herd or flock. This allows for the diversity of the small group to be maintained and for the breeder to gain expertise in the management of breeding animals. Other contributions of time, including professional skills and resources, can be made directly to the breed association. It is always of the greatest importance that breeders work to strengthen their breed's network, regardless of the flock or herd size they keep.

A Conservation Program for Two Parallel Herds

If numbers and space allow, or if the number of females is larger than can be served by a single male, or if there are several bloodlines to be conserved, a herd can be split into two parallel herds for conservation breeding. This strategy also works when two herds in different locations or under different ownership can be managed together.

A program for four or more distinct bloodlines can place these into two parallel herds instead of a single herd. This is usually a much better way to conserve all four than is running them sequentially through a single herd. If the latter is done, the males are fairly old before being used, and this increases the risk of loss of a male (and subsequently the line).

The parallel herd program is in some ways simpler than the single herd program. It does, however, require the management of two separate herds and the close cooperation among breeders. Each of the individual herds must be large enough to assure the production of good replacement males for each line, with some room for selection for production and other characteristics. This size is usually in the neighborhood of thirty females per herd, just as it was in the single herd program. This is especially true if selection for production will be practiced. The parallel herd idea can also work well with fewer females, and it can be a very good option for breeders whose herd or flock size is limited.

Sheep, such as these St. Croix, can be managed in parallel flocks.

The basic elements are the same as those of the single herd program described above. Each herd is divided into bloodlines. Single males from each bloodline will be used sequentially in each herd, serving to maintain the genetic distance between lines and allowing for the alternation of inbred and outbred conditions.

The most notable management difference in the parallel herd program is that each of the herds produces males for the other. All female offspring are retained in the herd of birth, but male offspring selected as sires are used in the other herd. That is, Herd 1 produces its own female replacements and replacement males for Herd 2. Herd 2 produces its own female replacements and replacement males for Herd 1.

The method of classifying individuals in parallel herds is also different than that for single herd program. All offspring must have a number to identify their herd, as well as a letter to identify their bloodline. Offspring are assigned the number of the herd in which they are born, that is, their mother's herd number. They are assigned the letter of their father's bloodline. At the same time, pedigrees are kept on all animals so that information on the female lines is always available.

The conservation program for parallel herds using cattle as the example is described in the text and illustrated in Figure 9.4.

The first step of the program is to divide the animals into two herds numbered 1 and 2 and three bloodlines A, B, and C. Herd 1 includes 1A, 1B, and 1C, and herd 2 includes 2A, 2B, and 2C. Three lines per herd are a minimum, since the program will work poorly with just two lines per herd.

The division into herds and lines can be accomplished several ways, depending upon the starting point. If three bloodlines already exist in a single herd, they can easily be divided into two new herds. If four, five or six bloodlines exist, some lines are put entirely into one herd or the other, and the remaining line(s) are divided between the herds. Generally, the lines which have larger numbers will be split. It does not make a significant genetic difference whether 1A and 2A represent the same line or are two different lines that are being managed together as a pair.

An advantage of the parallel herd program is that the lowered risk of inbreeding reduces the need to constrain the population as tightly as in the single herd program. One constraint is advisable, however. If numbers are large enough, it is preferable to select replacement bulls from among the offspring of cows from the same line as the sire. That is, the B replacement bull would have to be sired by the 2B bull and out of one of the 1B cows, rather than by one of the 1A or 1C cows. This assures a more even contribution of genetic material into the next generation, since the bull used was not only sired by a bull of the line but also has a maternal grandsire of the line. This is a subtle refinement, and is desirable though not essential.

Figure 9.4 The ALBC conservation program with two parallel herds of cattle

	Herd 1		Herd 2	
Cycle	**Bull used**	**Calves produced**	**Bull used**	**Calves produced**
1	2A	1A, because they are born in herd 1 and sired by bull A. Save a bull calf from a 1A cow, save good heifers from all cows.	1A	2A, because they are born in herd 2 and sired by bull A. Save a bull calf from a 2A cow, save good heifers from all cows.
2	2B	1B (born in herd 1, sired by bull B). Save a bull calf from a 1B cow, save good heifers from all cows.	1B	2B (born in herd 2, sired by bull B). Save a bull calf from a 2B cow, save good heifers from all cows.
3	2C	1C (born in herd 1, sired by bull C). Save a bull calf from a 1C cow, save good heifers from all cows.	1C	2C (born in herd 2, sired by bull C). Save a bull calf from a 2C cow, save heifers from all cows.
4	2A	1A, (born in herd 1, sired by bull A). Save a bull calf from a 1A cow, save good heifers from all cows.	1A	2A, (born in herd 2, sired by bull A). Save a bull calf from a 2A cow, save good heifers from all cows.
5	2B	1B (born in herd 1, sired by bull B). Save a bull calf from a 1B cow, save good heifers from all cows.	1B	2B (born in herd 2, sired by bull B). Save a bull calf from a 2B cow, save good heifers from all cows.

Of critical importance is the conservation of distinct lines, which serves to control inbreeding and maintain diversity.

The protocol outlined provides a basis for design of specific conservation programs for a conservation herd or flock. Every breed, every farm, and every breeder will have specific characteristics that may suggest useful deviations. The critical concept is the manner in which genetic diversity is maintained through the conservation of distinct lines within each herd or flock. The rotation of sire lines every breeding cycle through the entire population controls inbreeding and protects the genetic distance between the lines that is necessary for outcrossing. Conservation programs which replicate this strategy are likely to succeed.

Application to Other Species

The conservation breeding programs as described for a single herd or for parallel herds of cattle can easily be applied to sheep, goats, and swine. There are some differences, as these species mature earlier, have shorter gestations and produce more offspring per cycle. Multiple births are common for sheep and goats, and swine produce litters of young. The larger number of animals born each cycle gives breeders the luxury of being more selective in their choice of breeding stock as well as increasing herd or flock numbers more quickly. The programs can also be applied to horses and donkeys, though several modifications must be made.

Sheep and goats

Some breeds of sheep and goats are strongly seasonal breeders, meaning that the females are fertile for only a limited time of the year. For these breeds, the cattle example can be used exactly as it is. There is still only one breeding cycle per year, even though the cycle itself is shorter.

Other breeds are fertile for longer periods of time; this is referred to as having an extended breeding season or breeding out of season. Accelerated production systems have been developed to take advantage of this characteristic. One of these, the STAR system, mates a flock five times a year, with a portion of the females bred each time. It is possible to combine the STAR program with the conservation program by using five breeding cycles per year and a different sire line in each cycle. Three or four lines (or even six or seven) can thus be maintained while the flock is still managed as a single

Breeding plans for Myotonic goats can take advantage of nonseasonal reproduction.

unit. Use of exactly five lines is not recommended. If each line is mated at the same time every year, then not all lines would be under selection for nonseasonal breeding. In contrast, if three, four, six, or seven lines are rotated through the flock, each sire line will be used in every mating season over a period of a few years. One complication with this method is the number of animals necessary; the flock must always be large enough so that male and female replacements are produced in all five breeding cycles.

Swine

The conservation program is shortened for swine, with more breeding cycles per year. This is possible because swine mature early and breed year around. For example, boars might be used at about a year old, instead of two years old as in the cattle example. Though herd sizes are usually smaller than for other species, this is generally offset by the litters of young produced each cycle. Use of the conservation breeding program allows even a small population of pigs to be an effective conservation unit.

The number of sows and boars necessary will vary, but the consequences of inbreeding in small populations are still unavoidable, even in species that produce litters. As a result, most swine programs will require cooperation among several breeders. An advantage of such a network, however, is that the ratio of boars to sows used for breeding can be increased over that which is usual for most species. A program of six sows and three boars representing different lines (the boars being of different ages from very young to 18 months old) can work very practically on most farms, and it allows for some of the advantages of the linebreeding and linecrossing aspect of the conservation breeding program.

Horses and donkeys

The conservation breeding program must be lengthened and modified to fit the characteristics of horses and donkeys. These species mature more slowly than the others, so animals are older when first bred and each breeding cycle takes longer to produce replacements. Stallions and jacks can be difficult to manage, and few breeders have the resources to keep sires, much less multiple sires. Horses and donkeys form strong bonds with their pasture mates and for this reason it is preferable to manage them as relatively stable social groups, unlike the constant switching of males that is suggested for the other species. The investment in training necessary for stallions and jacks to be handled makes the rapid turnover of these males undesirable. An advantage for horse and donkey breeders is that the females have longer reproductive lives and can over time produce offspring from several different sires.

The management of horses and donkeys will, therefore, use the strategy of pair mating of individual males and females to maintain three or more distinct bloodlines within the herd. Linebred replacement males and both linebred and linecross females must be regularly produced. Linebred colts are evaluated as possible replacement sires, since each sire must be replaced by his linebred son. Linecross colts have less potential for conservation breeding, unless the breed is relatively numerous. They may, as geldings, become important ambassadors for the breed. Linebred and linecross fillies are retained for breeding, since both have a role to play in a conservation program.

There are a few different ways to accomplish conservation goals, depending on the management chosen for the herd. Horses, and to a lesser extent donkeys, can be bred in a pasture setting in which a stallion is kept with a small group of mares. The herd can be kept together either for the breeding season or for the whole year. The personalities of stallions vary, and some will insist on a relatively constant group of mares with few introductions or subtractions. Other stallions are less demanding, and new mares can be more easily introduced.

These Pryor Mountain Spanish Mustangs are managed in herds.

If a stable herd is desired for management considerations, or dictated by the personality of the stallion, then it is important to carefully assemble the females in relation to the male. The herd should include some females of the same bloodline as the stallion; they will produce linebred offspring. The herd will also include females from other lines to produce linecross offspring. The result will be a series of foal crops that include the necessary linebred and linecross replacements.

If the herd of mares or jennies can be shuffled over the course of years, with females added as needed, it is less critical to assemble a group at the beginning that comprises both linebred and linecrossed matings. What is important is to assure that both types of matings occur over a period of years.

Each foal in a conservation herd is a member of one of the bloodlines.

Some breeders use a modified form of pasture breeding by managing the females together as a herd for the entire year and putting a male into the herd only for the mating season. This strategy allows for use of different sires in the herd every year. If this strategy is adopted, and if the number of females is large enough, then it is possible to manage a horse or donkey program much like the programs recommended for the other species. Most of the time, however, herd sizes of horses and donkeys are much smaller than the recommended numbers for other species. With fewer females in the herd, the breeder will often have to use a single male for several cycles, either consecutively or every other year, until enough female replacements and a good male replacement have been produced.

More common than pasture breeding of horses and donkeys is the hand mating of an individual mare/jenny to an individual stallion/jack. ("Hand mating" is the mating of a female to a male during the period of time that she is in heat. The animals can be haltered and held for the mating to take place or they can be turned into a paddock together.) Handmating is the general rule for donkey reproduction, since many jacks are too rough on the jennies for pasture breeding. An advantage of hand mating is that females and males may be bred to produce the specific replacements needed, and several males may be used during a single breeding season. Over the course of a few years, an individual female can be mated to different lines of males.

The maintenance of multiple males, therefore, becomes the critical but limiting factor in horse and donkey conservation breeding. Because breeding cycles are longer, each sire may be used for several years of breeding in order to produce the necessary replacements. The difficulty of maintaining these sires, however, can be imposing. For this reason, a cooperating network of equine breeders is even more essential than with the other species. The network structure will maximize the number of sires used for breeding and spread out the responsibility for their management. In this way, breeders can make sure that sires of all lines are always available. This effort provides significant genetic returns in maximizing the use of male parents used for breeding each generation.

Introduction of New Breeding Stock

In all but the most tightly closed herds and flocks, new breeding stock will be introduced from time to time. How frequently animals are added depends upon the status of the herd and on the breeder's goals.

New stock is often added to reduce problems associated with inbreeding. The ALBC conservation program reduces the impact of inbreeding, though after five or six breeding cycles the male lines may become somewhat related. Inbreeding depression (signified by a decrease in viability or reproductive fitness) may occur when inbreeding coefficients reach the neighborhood of .3 (30%) or more. This eventuality can be avoided by the judicious introduction of a new breeding individual after two complete rotations of the three bloodlines (six breeding cycles). At that point, or when the breeder desires to boost certain traits, it can be useful to introduce a new male, female, or small group into the breeding population.

The introduction of a female is one way to bring in new genetics. This is a Karakul ewe and lamb.

The reason for the introduction determines the type of stock best introduced. If the breeder desires to increase performance for a characteristic in the herd, a male that excels in the desired characteristic may be added. This new male is used instead of one of the existing males, probably the one that is most deficient in the trait for which improvement is sought. A new line is created, combining the existing line (still represented by the females) and the introduced male. This strategy should strengthen performance over the entire herd for the next few generations.

If, in contrast, a breeder desires to reduce inbreeding but does not want to change the characteristics of the herd, then the addition of a female is more useful. The new female is added to one of the bloodlines, and a son of hers will be used as the line's next replacement sire. This young sire is half of the existing line (from his sire) and half from the introduced line (his dam). His offspring, in turn, are only one-fourth the influence of the introduced line. This strategy is usually sufficient to boost vigor, but it brings in less genetic material than if a male had been introduced. The existing line is kept largely intact.

The conservation of rare breeds, such as these Shetland sheep, can be challenging but also very satisfying.

A third option is to introduce a completely new bloodline to replace an existing line. This strategy requires the introduction of a group of animals. It is similar to the introduction of a male, but it is more broadly based and brings much more new genetic material into the herd. The introduction of a new line is used when it is necessary to greatly renew vigor (though this should never occur if animals are closely monitored). It is also appropriate if a new line of the breed has been discovered and must be incorporated into an existing herd to be conserved. The existing line that has been replaced may be combined with another one or retired, depending on the situation.

Disadvantages of the Conservation Program

A distinct disadvantage of the ALBC conservation program is that each individual sire has a relatively short breeding career, and this can be unsatisfying for breeders who desire to develop and use a superior male over many years. The use of a single sire for more than a season or two, however, allows him to become a genetic bottleneck, greatly diminishing the conservation value of the herd or flock. Using a male for more than one year also means that another male is not being used, so that all of the offspring produced are stacking up into one bloodline instead of being divided over several lines.

The second disadvantage is that the selection of all breeding animals must be done in a meticulous fashion to best benefit the whole herd or flock. A breeder will save and use a moderately inbred sire even when an outbred sire is more phenotypically desirable. This is because the moderately inbred male will generally outperform the outbred male in respect to the quality of offspring produced. The use of linebred sires (as opposed to linecross sires) also assures that the lines within the herd remain more distantly related over time. Though this close constraint of breeding animals can be a disadvantage, there is the opportunity to produce many excellent sires and sire prospects for sale to other breeders.

A third and more general disadvantage is that conservation breeding requires the breeder to break away from the usual practices and rewards of animal breeding and instead draw satisfaction from working for the long term good of a rare breed. The program is fairly strict, and the only way to succeed with small populations is to persevere with minimal deviations. In some situations, and for some people, this can be difficult to do.

San Clemente goat

Chapter 10. The Rescue of Livestock Populations with Very Small Numbers

The careful breeding of a rescue program can often protect very small populations from extinction. The goals of a rescue program are to increase numbers while maintaining genetic integrity. Rescue will take precedence over selection concerns until numbers have been increased. After several generations, management can shift to the regular conservation program.

It is usually the case that the population of interest (that is, the herd, bloodline, or even breed) in need of rescue has been reduced to a single "founder" male and a few "founder" females. Most of the time, a single male has been used to breed the females for several generations, his son has replaced him, and so on, resulting in an inbred population. The fact that the herd exists at all is some indication of resistance to inbreeding depression, but a rescue strategy is still needed to maximize genetic diversity and minimize future inbreeding to the extent possible.

The first step in any rescue program is to immediately collect semen from the single founder male. Collection of semen ensures that the male line can survive and the females can be bred pure even if the male himself is lost. The second step is to begin a breeding program immediately.

The rescue program begins with an evaluation of the population. Pedigree information describing the relatedness of individuals in the herd is very useful, but it is not usually available. Populations tended carefully enough to have pedigree

Leicester Longwool sheep in America are being conserved by a small network of breeders.

information are less likely to have been reduced to the crisis of minimal numbers in the first place.

Lacking pedigree information, some general assumptions can be made about the relationships between the females and the male. Females that are older than the male are obviously not his daughters, and they are probably not his sisters either. Rather, one is likely his mother and the rest are likely his aunts. Females that are about the same age as the male are likely to be his half-sisters. The females younger than the male are potentially his daughters, though their ages should be estimated and compared to be certain. The relationships between the females will probably be unknown, though the same assumptions can be made; that is, the younger females are likely daughters and granddaughters of the older females.

Without a deliberate strategy, the founder male will be overrepresented in the genetic makeup of all future generations. He represents a genetic bottleneck, and it is important to breed away from this bottleneck as quickly as possible. A short bottleneck is far less destructive of genetic structure than is a longer one.

Next, consider what resources exist in the herd. If females are bred to the male, the older founder females will produce offspring that are the most distantly related to him. The females the same age as the male, assuming they are his half-sisters, produce animals that are more related to him. The younger females, assuming they are his daughters, produce animals that are the most closely related to him. The only way to decrease the genetic influence of the founder male is to increase the relative genetic influence of the older founder females, who are the ones least related to him. This has to be done consciously and deliberately.

The goal, as for any population, is ultimately to have available at least three, and preferably more, relatively unrelated male bloodlines. The number of lines will vary, but three is a minimum and four to six is much more comfortable. Development of these male lines is the top priority early in the rescue program. In the absence of pedigree data, the surest strategy is to use the older founder females to develop these lines. If pedigree data are available, these recommendations can be modified to use the females that are most distantly related to the founder male in place of the older females.

A herd of cattle will be used to illustrate the rescue program, and this example is illustrated in Figure 10.1. Included in the figure are both years and generations, and these are not the same thing. All offspring of an animal, regardless the year born, are considered to be in the same generation. Calves produced by the same sire or dam over three years are the same generation genetically, even though they are of different ages.

Randall Blue Lineback cattle, once reduced to a single small herd, are being conserved with a rescue breeding program.

Individual animals are designated by a letter, and each letter stands for a founder animal, not a bloodline, as in earlier examples. Only the bulls, the older cows, and their sons are illustrated in the example since they are most critical to the program.

Founder generation: The first or founder generation is the starting herd. The herd in this example consists of a single founder bull, eight older founder cows, and ten or more younger cows. Since the founder bull is the only male available, he is bred to all of the cows. In addition, the bull's semen is collected and put into storage in the event that he is lost before the next generation bulls are ready to be bred. The founder bull is kept in the herd for additional insurance until the next generation is ready for mating. This insurance is worthwhile in truly critical projects.

After the calves from this first mating are born, and the cows come back into heat, they are bred back to the founder male, as he is still the only male available that is of breeding age. This provides additional individuals of the second generation.

Second generation: The second generation of the rescue program is made up of the offspring of the founder generation. All of these calves were sired by the single founder bull, either in cycle 1 or cycle 2.

The sons produced by the older founder females are of most importance. All of these

One of the goals of a rescue program is to increase numbers. These are Gulf Coast Native sheep.

bulls are retained for breeding, with selection made for soundness if numbers allow. They can be called the "half-brother bulls," as they are half-brothers to one another. The female calves produced by the older cows are also important as future replacements for their mothers. Even the female calves of the younger cows are of some value as they are increasing the overall numbers of the herd. The only animals not retained for breeding are the bull calves from the younger females, since these calves are likely to be more closely related to the founder bull than are the bull calves from the older females.

When the half-brother bulls are mature, all of them will be used for breeding. The older group of half-brothers will be mature at about one year old, in year 3. The younger group of half-brothers will be mature in year 4.

In order to use all of these young bulls for breeding, the cows are divided into small groups. Each group should have some older cows and some younger cows, with daughters to be included in the same group as their dams (where this relationship is known). There should be the same number of cow groups as there are young bulls, and a young bull is mated to each group. The critical element of this step is the mating of each of the older founder females to a young bull. In most cases, the bull will not be mated to a group that includes his mother, though there are some exceptions when such close inbreeding would be chosen to concentrate the genes of one founder female.

The reason for using all these young bulls at the same time is that the clock is ticking on the older cows, and the program is far from finished with them. The program has grown from a single, original sire to include three or more young sires. These sires are all half brothers, but a start has been made.

At this point it is useful to review the relationships between the young bulls and the cows in the herd. The older females are relatively unrelated to the young males; except for the bulls' respective mothers, these females are their great aunts of some sort. The females of intermediate age (that is, probably sisters of the founder male) are aunts to the young bulls and therefore somewhat related. The youngest original females are half-sisters of the young bulls, as they share the same father.

Third generation: The third generation is made up of the offspring of the half-brother bulls. Again, the bull calves out of the older founder females are especially important and all of them will be used for breeding. These "quarter-brother bulls" are only one-fourth the influence of the founder bull and thus only one-fourth related to one another.

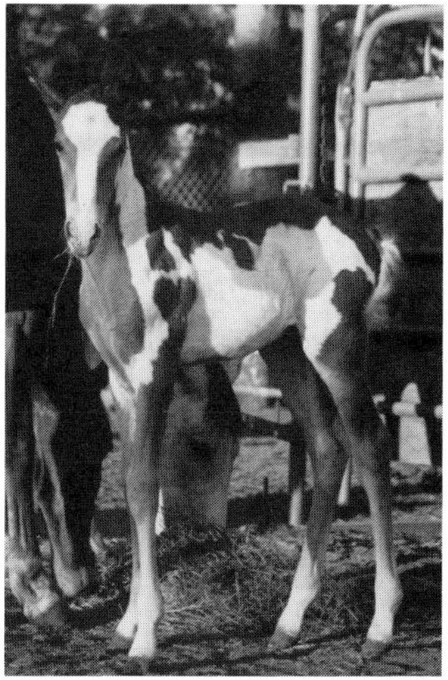

Rescue of rare horse breeds, such as the Spanish Mustang, is a slow process due to horses' long generation interval.

As before, the young bulls are used for breeding all at one time across the herd. Though the program works best with four young bulls, there may be three or five available. All of them should be bred, and the females are re-grouped if necessary so that there is one group of cows for each bull. Each group should still contain one or more older foundation cows as well as younger cows, and in general the young bulls are not mated to their mothers. This is roughly equivalent to rotating the males to a new female group. The mating of several small groups can be a management nightmare, but this needs to be done if the rescue is to be effective.

When the offspring of these matings are born, one can say that the matings of the older cows and the young bulls have established distinct bloodlines. Each line is based on an older founder cow and includes the male and female descendants of that cow. In Figure 10.1, eight distinct lines have been created. The genetic influence of the older founder cows has been successfully increased relative to that of the founder male. This is done through the sons of the older cows, who can more easily make genetic contributions to the entire herd.

The description of the program has concentrated on the male offspring, since they are usually responsible for the bottleneck. In every generation female offspring are also born, and they are managed in the same groups with their mothers. Eventually, each founder cow has not only provided sons for mating with other lines, but she has also surrounded herself with daughters by different sires. These daughters are part of her line, and when the rescue is complete they can be managed as a bloodline in the conservation breeding program.

At this point, or after one more generation, it is possible to shift from the rescue program to the conservation program, since multiple lines now exist.

Fourth generation and beyond: If the older cows are still sound it is possible to do essentially the same thing again. The older cows' bull calves of the fourth generation are "eighth-brother bulls," only one-eighth the influence of the founder male and only one-eighth related to one another (at least as far as can be documented by records from the program). With one more cycle, the relatedness is only one-sixteenth. It is advisable to keep breeding the older females in this manner as long as they are healthy.

The rescue plan is straightforward in theory, but in practice it demands many years of work. In year one, the original male is used. The first new bulls are born almost a year later and they are not ready to breed until year three. Their sons are born in year four, and they are ready to breed in year five. The sons from these are born in year six (these are the eighth-brothers), and only then is the point reached where the relatedness between bulls becomes small enough that distinct lines have been created. With sheep, goats, pigs and poultry the number of years is fewer, but with horses or donkeys it is even greater.

Complementing the management of live animals is the collection of semen. A frozen store of semen is a critically important insurance policy, and each sire should be represented in semen storage as soon as he is mature. The founder bull is collected at the beginning of the rescue. Saving semen from the first crop of half-brothers is not essential, but desirable. The next generation (quarter brothers) should certainly be collected, and the next (eighth brothers) is even more useful. It is always possible to lose a male through disease or accident, and if a loss occurs, that line is either threatened or extinct—unless semen is in storage. It is especially wise to freeze semen from a male whose dam has died, since she can no longer contribute except through him. The semen can be used in the event that a sire line become extinct through accident or oversight.

Though it is relatively easy to have cattle semen collected and frozen, the same may not be true for the other livestock species. The possibility, however, should be investigated and pursued if feasible.

Figure 10.1 The ALBC rescue program for a very small population of cattle, showing the use of older females and younger males

Generation	Year	Bull used	Older founder cows bred	Young bulls produced (estimated according to the law of averages)
First	1	A	B, C, D, E, F, G, H, J	AB, AD, AF, AH
	2	A	B, C, D, E, F, G, H, J	AC, AE, AG, AJ
Second	3	AB	D, E	AB/E
		AD	F, G	AD/G
		AF	H, J	AF/J
		AH	B, C	AH/C
	4	AC	D, E	AC/D
		AE	F, G	AE/F
		AG	H, J	AG/H
		AJ	B, C	AJ/B
Third	5	AB/E	F, G	[AB/E]/F
		AD/G	H, J	[AD/G]/H
		AF/J	B, C	[AF/J]/B
		AH/C	D, E	[AH/C]/D
	6	AC/D	F, G	[AC/D]/G
		AE/F	H, J	[AE/F]J
		AG/H	B, C	[AG/H]/C
		AJ/B	D, E	[AJ/B]/E

Pilgrim geese

Chapter 11. The ALBC Conservation Breeding Program for Poultry

The poultry species are among the smallest of the domestic livestock, and their husbandry is not difficult. The characteristics of poultry, however, make breeding for genetic conservation more technically complicated than that of the other species.

The conservation of poultry has the group, rather than the individual, as the unit of interest. The status of the flock is of even greater importance in birds than in mammals since it takes more effort to maintain individual pedigree information. This includes the matching of an individual to its eggs, and the individual egg to its specific resulting youngster. Relatively large numbers of birds must be kept, and reproduction is generally done annually given the short reproductive life of these species. In addition, all species of poultry are highly vulnerable to predation. For these reasons, poultry conservation is made far easier if done by a network of several breeders.

Conservation programs for chickens, ducks, and turkeys are similar. These birds can be mated in small groups, with sires easily rotated among the groups. Conservation of geese is somewhat different, because geese form pair bonds and live in stable groups of one male with one to three females.

Poultry social structure is an intricate system of relationships among the members of a flock. Birds have tightly ordered hierarchies, and efforts should be made to assure that once a group is formed there are few changes in its makeup. This is especially true within sexes, so that new males should not be put in with established males, nor new females introduced into a flock of females. The introduction of new birds will often result in competition and fighting to reestablish social order, increasing stress and reducing fertility. Males can, however, be rotated through a flock of females with little social upheaval, and the use of this rotation will be described below.

It is generally true of poultry breeding systems that the parents are retired when the next generation matures. It is rare for generations to overlap in poultry breeding. The usual generational interval, and at least for chickens and turkeys, is one year. Offspring are generally brooded in a separate flock until maturity, at which time they replace their parents in the breeding program.

Elements of a Poultry Conservation Program

The objectives of a poultry conservation program are the same as those for a livestock program: the management of an individual flock so that genetic diversity is maintained and the flock resource best serves the conservation of the breed. In general, poultry genetic diversity is protected through maximizing the use of male parents in each generation, randomizing breeding, and selection for soundness and breed characteristics.

Conservation of poultry breeds, such as the New Hampshire, is technically challenging.

Production traits are an important part of a breed's genetic heritage, so selection for production should be an element of conservation programs for all but the very rarest breeds. In fact, it was emphasis on specific breed points alone, often causing a reduction in production traits, that led to the split between commercial and exhibition poultry breeders early this century.

Identification: Central to a poultry conservation program is the need to identify individual birds, bloodlines, or groups of birds within the flock. This is generally done by a colored or numbered leg band, or by toe web punching. The identification of birds may be done to facilitate the management of different groups together in one large flock, or it may be done as part of management in separate pens.

As with livestock, the conservation breeding of poultry is made more complicated due to the need to use a relatively large number of males. The management of these males also accounts for much of the extra work of husbandry. Unlike livestock, however, birds are small in size. It can be far easier to manage several groups when one is talking about pens and cages rather than pastures and barns.

These Delaware chickens, like most poultry, are managed in a flock.

Housing: Well-designed housing can facilitate conservation breeding. A large pen or house with several small "apartments" for three to five birds, for example, can be useful to the breeder. It then becomes easy to manage the female birds as a single flock and divide males into several smaller groups which can be rotated in and out of the flock for breeding. The stability of these smaller groups reduces stress and protects the fertility of the birds.

Some breeders are able to pasture their birds outdoors during the day and keep them housed at night. With birds spread out on range, it may become possible to manage the entire group of males and females as a single flock year round. The dispersal of roosters among the hens and the space will reduce fighting somewhat and allow several males an opportunity to mate. Even so, more losses will occur on range due to fighting and predation, and beginning numbers should take this into account.

Ages and numbers: Poultry are fertile over a period of several months or even years, but for best results they should be reproduced at the age of maximum fertility and vigor. For chickens, ducks, and turkeys, that is approximately one year old. Geese are most fertile between the ages of two and five years, though they may be productive for longer than that.

It is customary in chicken and turkey production to use a single male for eight to ten females. For conservation purposes, such a bottleneck should be avoided. A ratio of one male to two, three or four females is desirable. Commercial ducks are kept at a ratio of one to six, though for conservation purposes ducks should be kept at a ratio of one male to three or four females. Geese, of course, insist on stable groups of one male to one to three females for the heavy breeds, and one male with up to five females for the Chinese breed.

These Bronze turkey poults resulted from successful incubation and hatching of eggs.

The use of multiple males for breeding is necessary because birds are only used for one cycle before being replaced. Each female should be mated to more than one male to assure the broadest genetic contribution to the next generation. The sequential use of small groups of males across an entire flock effectively breaks up a single breeding cycle into many smaller time periods, each with a different group of males.

The total number of birds in a conservation flock should be as large as the resources will allow. Ideally, twenty-five to fifty hens and twenty or more roosters should be used as parents for each generation. This "magic number" to minimize genetic losses in a sustainable flock has been suggested by poultry scientist R.D. Crawford of the University of Saskatchewan. A flock this large might be a burden on a single breeder, but it can also be shared among several breeders, who rotate males each year. More important than precise numbers, however, is the establishment of an effective breeding program.

The practice of splitting a large flock among several breeders has the additional and profound benefit of insuring against devastating predator or disease loss. The number of breeder flocks is a significant component of breed status in poultry, since a single flock of any size can be lost in a single night. Because of the threat of predators and other factors, poultry conservation breeders must expect some loss. In chickens and turkeys this loss is frequently as high as one percent per month and the numbers of birds at the beginning of a program should take this into account.

Incubation: Added to the theoretical aspects of breeding and selecting birds is the need for incubating and hatching eggs. This is as much an art as a science, and successful breeders use many different strategies. Some breeders choose to incubate eggs under foster hens, usually of a broody breed such the Silkie. Others use artificial incubators. Various methods of incubation are described in detail in many of the books dealing with poultry. Each species of poultry has its own peculiarities that must be understood if natural or artificial incubation is to be successful.

Some breeders have the opportunity to pay someone else to custom hatch their eggs, and contracting for this service is often worthwhile. Home incubators can be difficult to operate and it is frustrating to lose weeks of conservation work with an incubator problem, power failure or human error. It is essential to plan for successful incubation —even then, don't count chicks before they hatch.

Two Methods of Poultry Conservation

The use of relatively large sized flocks, multiple sires, and annual reproduction are necessary for poultry conservation. With these objectives in mind, two methods of genetic conservation of chickens, ducks, and turkeys are described below. The first program is likely to be more widely applicable, though the second program will be appropriate whenever there are distinct lines present in a flock and the breeder desires to maintain the genetic distinctiveness of each one.

Multiple sire mating with selection

The conservation goal of this program is to use a large proportion of males, multi-sire mating, and selection for soundness to maintain the genetic viability of the population against genetic losses from drift and inbreeding. The critical element is the use of many males in reproduction of each generation. Utilizing several small groups of males sequentially assures that many males have the opportunity to mate.

The following protocol for the maintenance of poultry conservation stocks was described by R.D. Crawford in 1987 (personal communication). This protocol also appears in *Raising Rare Breeds, Livestock and Poultry Conservation: A Producer's Guide* by Jy Chiperzak of Joywind Farm Rare Breeds Conservancy, Inc. The ALBC program represents a modification of Crawford's work by the authors and includes contributions from poultry scientist Robert Hawes of the University of Maine.

All females are kept in a single flock, which eases management and provides a stable social environment. The males are divided into three or more stable groups of three to five birds each. Keeping each group of males in its own pen when not with the females will reduce injury due to fighting. When the breeder is ready to begin reproduction, usually in the spring to early summer, then the small flocks of males will be rotated sequentially through the flock of females.

The first group of males is put in with the females. The breeder should wait for about

Identification of individual birds, like this Dominique fledgling, is an important element of poutry conservation.

a week to establish fertility, and then collect eggs, saving them for up to two weeks. Eggs can be stored for two weeks without losing fertility; saving eggs for one incubation eliminates several small hatches. The first batch of eggs is then ready to be incubated.

The first group of males is removed from the female flock and replaced with the second group of males. Eggs are not collected for a week, while the semen in the females' reproductive tract is cleared. Then, eggs are collected for two weeks. The spacing of incubation three weeks apart can also have a management advantage. Since chicken eggs hatch in three weeks, the same incubator can be put back into service to hatch the second group of eggs. (The other species take longer to hatch, but if the beginning of the second cycle is delayed for an additional week, the incubator can be used again.) The third group of males is managed likewise.

Using this method, each group of males makes up its own hatch. The disadvantage is that an incubator failure requires a repeat of the breeding cycle, though this can be done relatively easily. The advantage is that the chicks from each group of sires can be easily identified at hatching, since only one group is in the incubator at one time.

Upon hatching, birds can be identified by leg bands (which must be changed frequently to accommodate rapidly growing chicks) or by toe punching. Identification by group means that sires from each group are used for breeding the next year. This assures an even contribution from all male groups and prevents the loss of a male group from one generation to the next. Females can also be identified, and this provides equal representation of daughters from each group for the next year. Representation from all sire groups is especially important when total numbers are small. In this way, diversity is being maximized, even if these separate groups have not been isolated long enough to become separate bloodlines.

The breeder's goal is to hatch enough offspring so that birds for the next generation may undergo some selection based on health, soundness, vigor, and breed characteristics. An excess of good chicks can always be used to establish new conservation flocks. Once the new generation is in place, the parent generation is retired from breeding. It is desirable to exchange chicks, poults, or ducklings of both sexes, or just males, with other breeders each year or every other year. This avoids divergence of the flocks and inbreeding depression. Divergence of the flocks is not a goal. It is more useful in a conservation sense to manage small flocks cooperatively as one large conservation flock.

The poultry conservation protocol does not seek to develop or maintain individual lines of the flock. Most often, breeder flocks are begun with a group of hatchery chicks, which can be assumed to have come from a single hatch. For this reason, it is not particularly useful to separate the birds into lines since this cannot be done based on genetic relationships. Instead, the goal is to conserve the genetic diversity present at the beginning of the flock, and this is accomplished through the use of many males in rotating groups to randomize breeding. Individuals which exhibit unsoundness or deviations from breed type are culled.

Use of distinct lines

If a flock has originated from distinct foundations, the breeder has to decide whether to blend these lines together or to keep them separate. Blending the lines together—as described in the multi-sire breeding strategy above—will increase the health and production traits of the resulting flock, and selection can be used to maintain these qualities. This is appropriate when the breed to be conserved has several relatively weak, inbred lines, and the hybrid vigor generated can be used to reestablish production potential.

Replacements are hatched each generation for each bloodline. These are New Hampshire chicks.

Alternatively, a breeder may choose to maintain distinct bloodlines. This provides an opportunity to study performance and other differences and it keeps open the possibility of line crossing in the future. This approach is appropriate when the distinct lines available are themselves viable and productive, and it is similar to the livestock programs described earlier.

In this program, the birds are divided according to their bloodlines and then managed in separate pens. After reproduction, female offspring replace their mothers in the same pen, and male offspring are rotated to the next pen. Each line must be assured of raising replacements, so a minimum of five females are needed per pen, and ten is better. Figure 11.1 provides an illustration.

This program maintains the distinct genetics of each strain. It can produce some of the inbred, outbred effect described in the livestock program. A disadvantage of moving males in a circular rotation through all of the bloodlines is that eventually the flock becomes inbred in the same direction. This disadvantage, however, can be minimized if outcross males are available from another source every few generations.

A second management method designed to decrease inbreeding through the use of distinct lines is the management of eight lines distributed among two groups of four pens each. This is somewhat like the parallel herd program for livestock and is illustrated in Figure 11.2.

Birds are grouped according to their bloodlines. The lines are labelled A, B, C, D (Group 1), and E, F, G, and H (Group 2). For the first four years of the program, males

Figure 11.1 The ALBC poultry conservation program using four lines within one flock

Year	Female line used	Male line used	Offspring produced
1	A	A	A
	B	B	B
	C	C	C
	D	D	D
2	A	D	A
	B	A	B
	C	B	C
	D	C	D
Rotation continues in Year 3 and beyond.			

are rotated through pens in the same group, that is A to B, B to C, C to D, and D to A; in the second group, the rotation is E to F, F to G, G to H, and H to E. Each group is being managed like the four-line program described above.

After four years, the birds in all four pens of each of the two groups have become

Figure 11.2. The ALBC poultry conservation program using eight lines in two parallel flocks

Year 1	**Group 1**				**Group 2**			
Females	A	B	C	D	E	F	G	H
Males	A	B	C	D	E	F	G	H
Offspring produced	A	B	C	D	E	F	G	H
Year 2								
Females	A	B	C	D	E	F	G	H
Males	D	A	B	C	H	E	F	G
Offspring produced	A	B	C	D	E	F	G	H
Years 3 and 4, rotate males to the next pens within the same group								
Year 5								
Females	A	B	C	D	E	F	G	H
Males	E	F	G	H	A	B	C	D
Offspring produced	A	B	C	D	E	F	G	H
Years 6, 7, and 8, rotate males to the next pens within the same group (same pattern as Years 2-4)								
Year 9, exchange birds between groups as for Year 5								

moderately inbred. In the fifth year, the males are swapped between groups, that is, A and E males are exchanged; B and F males are exchanged; C and G males are exchanged; and D and H males are exchanged. This has the result of counteracting the inbreeding that has occurred and it also provides a periodic boost of outbreeding. The disadvantage of this method, of course, is the necessity of managing eight pens with close identification of birds.

There are alternatives to the pen management system, at least for part of the year. Birds may be banded or toe punched to identify their bloodline and then the females, or all the birds, managed together as a flock until reproduction begins. At that point, the males and females are sorted into pens according to their bloodlines. When reproduction is completed, the hens are put back together as a single flock while the chicks grow to maturity as replacements. Males cannot be reunited into one large flock after having been split into small groups (because of fighting), and they will need to be managed in their small groups until they are retired.

Pilgrims, like all geese, are bred in small stable groups.

The breeding programs described for poultry have some theoretical disadvantages when compared to the conservation program for the other species, mostly involving the eventual evenness of increase in the inbreeding coefficient across the flock. As long as multiple males are used in each of the breeding groups, however, the genetic impact of inbreeding is decreased.

Management of geese for conservation

Geese form stable pair bonds or groups of a single gander with two, three, or more geese. This means that they cannot be bred under the same conservation program suitable for chickens, ducks, and turkeys.

The easiest conservation system for geese rests in a network of individual flock owners who exchange goslings every generation or two. For most breeds, unsexed goslings rather than baby ganders will be exchanged, since goslings are difficult to sex. The Pilgrim, with males and females of different colors, is an exception. Pilgrim breeders will be able to separate baby geese and ganders and therefore would be able to swap baby ganders.

In this network of geese conservationists, each flock is the equivalent of a separate breeding pen, and the exchange of stock is somewhat like the rotation of males. Participation in a network, with several small groups of geese, allows breeders to maximize the number of male parents each generation. Geese present some of the same challenges for conservation management as do horses and donkeys!

Selection and Conservation

Selection of poultry for breed type, health, soundness, and production qualities should be done as part of breed conservation. Production qualities are essential to production breeds, and these must be conserved if the complete genetic heritage of the breed is to survive.

Birds are generally evaluated for health, soundness and breed characteristics at maturity, when meat traits are considered. Males to be used for breeding are identified, and culls can then be used for meat. Selection based on egg production, however, must wait until the females have been laying for some time. If birds are to be bred at one year old, egg production is evaluated in the spring, just prior to reproduction.

Poultry breeders should be conservative in any selection decisions that take place months before reproduction. The poultry species are highly vulnerable to predation, and birds can easily be lost. It is wise to keep extra birds in the flock as insurance until reproduction begins.

One "performance" trait of interest to some breeders is broodiness, which is the propensity to set on a clutch of eggs until hatching and to raise the offspring. This is a useful characteristic for those breeders who prefer to let the hens incubate eggs, but it does decrease egg numbers, and so it is not desired in egg production breeds.

Selection for meat: Selection of meat traits is relatively simple, since the performance of many individual birds is measured at a single point in time. The time chosen is one that is appropriate to the species and the breed: eight to sixteen weeks for chickens; ten to sixteen weeks in waterfowl; and sixteen to twenty-eight weeks for turkeys. At the time of evaluation, the birds which are superior in meat conformation, health, feathering, and soundness will be kept for breeding, and the remainder will be

Production breeds like this Light Brahma should be selected for environmental adaptation and performance characteristics.

used for meat. Body weight is the most obvious trait for evaluation but some subjective measure should also be made of breast width. Good legs are especially important in large-bodied meat birds, so it is important that only sound individuals be retained for breeding.

Rare breed meat birds will not grow nearly as quickly as industrial hybrids, such as the Cornish-Rock cross broiler, but they should exhibit much better soundness and health in extensive production systems than do the industrial birds. Unless breed numbers are critically low, it is appropriate for selection of these breeds to include an element of improvement, since they resulted from improvement breeding.

Selection for eggs: Selection for egg production is a more difficult management exercise than selection for meat production, because performance is to be measured over a period of time. Accurate records must kept on individual hens, which generally requires that they are banded with an individual ID number (as well as bloodline ID). Record keeping can be laborious, but is necessary if selection for eggs is a priority. Without selection, egg production will drop over the generations.

Selection for egg production has been done traditionally by trap nesting. A trap nest is a laying box with a door that springs shut when a bird goes inside; after laying an egg, the female is "trapped" inside until released. This means that the breeder has an opportunity to identify the individual and record that she did lay an egg as well as information on egg quality and size. An alternative is to put hens in individual pens at night and release them during the day after egg production has been recorded. The females that excel can then be used for reproducing the next generation of birds, and the less productive females can be culled. The production test is most accurate if it is conducted over several months. Longer length tests in general will correct for seasonal effects.

Another option is to measure egg production for five days a month, over several months, as this measure is well correlated with annual rates of lay. Complete production could also be measured for only one month, and this also is correlated with annual rates of lay, although it may be confounded by seasonal effects. An alternative to trap nesting or individual housing is to manually palpate the hens early in the morning to see which ones will lay eggs that day. This is a time consuming task, but may be as practical for some breeders as the alternatives.

Another option in selection for egg production is to divide females of each bloodline into two or three separate pens. The egg production of each pen can be monitored separately, and the most productive pen of each line can be used for the reproduction of that line. The fact that this is a group selection rather than an individual selection means that it is less powerful, but it still maintains and slowly improves the egg production of each line of birds.

Indian Runner ducks and other improved egg breeds are selected for laying ability as part of conservation.

The replacement flock can also be hatched from eggs that are laid out of laying season when the birds are over eighteen months old. The theory is that the birds producing eggs out of season are the more persistent layers, with higher overall production. This also has the advantage that while maximum selection for egg production might not be occurring, there should be little loss of existing potential for egg production. The disadvantage is the necessity of hatching the eggs out of season, when environmental conditions are less favorable and raising the young is more difficult.

There are external clues to relative egg production in chickens and ducks, and at a bare minimum, breeders should sort females based on these features. Those hens which are laying have brighter comb and skin color and greater width between the pelvic bones. Ducks that are laying regularly show bleached legs; even the green spots will be bleached. This can provide some idea of laying ability.

Use of some technique to measure and select for egg production in dual purpose breeds of chickens and ducks is as important as it is for the specialized egg producing breeds. It is also important for meat breeds of all poultry species to have some selection for egg-laying ability, since this can be a limiting factor in production.

Very Small Populations of Poultry

When a breed of poultry reaches a critically rare status, a rescue program can often prevent extinction. The principles of poultry rescue are similar to those discussed earlier for livestock.

The priority during a rescue program is to breed all birds that are available so that numbers are increased and genetic diversity present is maintained. Depending on the time of year, resources available, and numbers of birds, this can be done as in the multiple sire breeding program above, or in pens where individual males (or groups of males) are rotated through to breed with individual females or groups of females. The maximum number of different matings should be used, and the maximum number of eggs should be hatched. All young are retained for possible breeding.

During a rescue program, it is particularly important to establish several flocks so that the breed is protected against loss from predation or disease. It may also be helpful to enlist the help of other breeders in the hatching of eggs and brooding of offspring.

Depending on the status of the breed, some selection may be made for soundness or breed characteristics in the second or third generation. Since the population has been through a genetic bottleneck when it was critically rare, breeders should be careful to use the maximum number of male parents for the next several years.

Conclusion: A Call to Conservation

"In the end, we will conserve only what we love, we will love only what we understand, we will understand only what we have been taught." —Baba Dioum, Senegalese conservationist, quoted in E.O. Wilson's *The Diversity of Life*.

One hundred American breeds of domestic cattle, goats, horses, donkeys, sheep, swine, chickens, ducks, geese, and turkeys are in danger of extinction. These breeds were carefully crafted by dedicated breeders to fit a range of conditions, tasks, and needs. Their beauty, dignity, and usefulnesss were all celebrated and they were partners in both agriculture and culture.

What if these animals become extinct? The Earth will have lost a unique segment of its biodiversity. Agriculture will have lost the genetic resources critical for adaptation to changing conditions and methods of production. And future breeders will be deprived of the flash of a Spanish Mustang in the misty morning, the gentle clucking and crooning of the Dominique hen to her chicks, and the rich palette of colors in the Navajo-Churro sheep.

If these breeds are lost, the animal kingdom—and the human experience—will be diminished forever.

Unlike wild species, domestic breeds must be conserved by human beings. Domestic animals share habitat with humans, that habitat known as agriculture. It is the use and reproduction of rare breeds in agricultural settings that is essential for their conservation and best reflects our long heritage of interaction.

People and animals are meant to be together.

The conservation of rare breeds presents several challenges in the biology and management of the domestic species themselves, as well as those obstacles imposed by the existing agricultural systems and networks. These challenges must be acknowledged and overcome for conservation to succeed. An individual steward with an understanding of conservation breeding and the husbandry skills to manage animals, however, can make a substantial contribution to breed conservation.

The conservation of rare domestic breeds can be done in one's own backyard. The work is difficult, but it does not have to be expensive nor take a lot of space. But it will not wait. Join the American Livestock Breeds Conservancy in this effort and find the unique rewards of protecting rare breeds from extinction.

Glossary

Adaptation—the various modifications of an animal or population to better fit the conditions of its environment.

Artificial insemination—the collection of semen from males, processing and storage of the semen, and its placement within the body of the female to fertilize ova and produce offspring.

Biodiversity—the overall variety of habitats, species, and genes which allows biological systems to adapt and evolve.

Bloodline—a group of animals within a breed that are more closely related to each other than to the breed as a whole; also called strain.

Bloodtype—a documentation of the blood cell and blood protein characteristics of an animal. Bloodtypes are frequently used to verify parentage and they are sometimes used to investigate the relationships of breeds to one another.

Breed—a group of animals that are similar enough to be logically grouped together, are distinct from others of the same species, and when mated together will reproduce this distinguishing type.

Breed association—an organization of breeders and other enthusiasts on behalf of a specific breed. The activities of a breed association generally include the development of policies regarding registration, operation of a registry, and promotion of the breed.

Breed registry—a record of the individual animals which are members of a specific breed. In most breed registries, each individual is assigned a registration number and is described by its name, date of birth, pedigree, identification (tattoo, brand, eartag), color and other phenotypic characteristics, breeder, and owner.

Breed standard—the definition of a breed, including a description of its phenotypic and performance characteristics. In general, individual animals must conform to the breed standard to be registered, and those which deviate are eliminated from the breed. The standard serves as a goal for those who seek to breed animals true to the breed's history.

Broodiness—the propensity of a female bird to sit on a nest until eggs are hatched and then to raise the offspring. Such a female is called "broody."

Closed herdbook—the registry policy which limits registration to only those animals which are offspring of registered parents.

Conservation—the protection of a resource to prevent its loss through overexploitation, neglect, or destruction. The term as used in this book describes the protection of living populations of animals while they continue to adapt to the environment.

Conservation breeding—the management and reproduction of a herd or flock of animals as a genetic resource to be protected for the future. The objective is to maintain the original genetic identity of the breed by using selection procedures that are consistent with the breed's history.

Crossbreeding—the mating of animals of different breeds, usually done to take advantage of hybrid vigor or to produce offspring with a potentially useful blend of traits from the parents.

Cryopreservation—the freezing and storage of genetic materials, particularly semen and embryos, for future use.

Culling—the removal of undesirable individuals from the breeding group. Culling can also be defined more generally as the selection of breeding animals from a group.

DNA fingerprinting—the identification of an individual, a breed, or other group of animals by comparison of sequences of genes and genetic markers to those of other individuals or groups of animals. This process is used to establish the parentage or breed of an individual, or it may be used to establish the relationships between breeds or groups of individuals.

Dam—the female parent of an animal.

Domestic—animals which live in or near the habitations of human beings, with the implication that husbandry, selection, and reproduction is under the control of humans.

Drift—the random loss of genetic variants from a population, especially true for genes that occur at low frequencies in small populations. Eventually, no individual in a generation will have inherited the rare gene, at which point it is gone from the population. Drift is nondirectional and is not caused by selection; also referred to as genetic drift.

Dual purpose—animals that are selected for performance in two or more characteristics. Dual purpose cattle, for example, have been selected for useful levels of production of both meat and milk, while dual purpose chickens produce both eggs and meat.

Effective population size—the mathematical calculation used to estimate the genetic size of a population, often illustrating the importance of maintaining male as well as female numbers to avoid inbreeding.

***Ex situ* conservation**—the conservation of a breed or population outside of the habitat from which it originated and to which it is adapted. This is particularly applicable to feral stocks, which are often removed from a habitat to facilitate the conservation of native plants and animals. *Ex situ* is in contrast to *in situ*, which means within the habitat of origin.

Extinct—no longer in existence. A breed may become extinct when its genetic integrity can no longer be documented, or when the breeding group is so small as to no longer be viable. Breeds may become extinct even as the total numbers within the entire species remain high. When breeds become extinct, the genetic range of the species may be permanently narrowed.

Fecundity—the capability of having many offspring; prolificacy. When applied to livestock, this term generally means having a relatively large number of offspring at one time, such as the Finnsheep, which may have four lambs at once.

Fertility—the capability of producing offspring.

Feral—a domestic animal or population which has reverted to the wild and is, as a result, under natural rather than human selection. A feral population is a reproducing group of feral animals.

Flockbook—the published record of animals which have been registered in a specific breed. Information is listed by registration number, and it includes name, pedigree and date of birth. The term herdbook is used for breeds of cattle, goats, and swine; studbook for horses and donkeys; and flockbook for breeds of sheep.

Forage—food for animals which is taken by browsing or grazing. Grass and legume pasture is the most commonly used forage in America. Forage utilization measures the ability to thrive on a forage diet and may include several elements, including the motivation to graze aggressively and the ability to digest forage efficiently. The term forage is also used as a verb, in which case it means the activities of animals as they graze and browse on their own.

Founder—an animal that originates a population.

Founder effect—the permanent shaping of identity and type in a population as a consequence of the identity and type of the animals which founded the population. When no other animals are introduced, the only types possible are those represented by the founders.

Gene—the unit of the code of inheritance. Each gene is a discrete bit of information that is ultimately reflected in the traits (appearance, behavior, and performance) of an animal.

Gene pool—the sum total of the genetic variation available within a population.

Genetic drift—the random loss of genetic variants from a population, especially true for genes that occur at low frequencies in small populations. Eventually, no individual in a generation will have inherited the rare gene, at which point it is gone from the population. Genetic drift is nondirectional and is not caused by selection. Also referred to as drift.

Genotype—the genetic type of an individual or breed which determines its appearance, behavior, and performance. The consistency of genotype in pure breeds is reflected in the consistency of appearance, or phenotype.

Genome—the total genetic material in an individual, breed, or species.

Genetic bottleneck—the effect on a population when numbers are greatly reduced. When only a few animals are the basis for future generations, genetic variation is often eliminated from the population. Genetic bottlenecks may be short (such as a single generation) or long (several generations).

Genetic diversity—the presence of a large number of genetic variants within a breed, population or species of animals, allowing for adaptation to the changing conditions of selection.

Genetic erosion—the loss of genetic variation within a breed, population, or species. This usually happens through loss of numbers, although it can happen through loss of variability even if numbers remain high.

Genetic uniformity—the presence in a breed, population, or species of very few genetic variants for each characteristic. This situation is often created by strong selection (human or natural) for a specific environment. A uniform population, lacking variation, has less ability to adapt to changing conditions than does a variable population.

Genetic variability—the presence in a breed, population, or species of large numbers of genetic variants for each characteristic. This situation results from relatively relaxed selection (human or natural) for many traits. Genetically variable populations have more success in adapting to change than do uniform populations.

Gestation—the length of time of pregnancy, between conception and birth of offspring.

Grading up—a registry policy designed to expand the numerical base of a breed. In most grading up programs, purebred registered males are bred to selected females of another breed; the female offspring of this mating are then bred to purebred, registered males. Their female offspring, in turn, are also bred to purebred, registered males. When female offspring reach 3/4 or 15/16 or 31/32 purebred (depending on the registry policy), they may be entered into the registry as purebred.

Habitat—the environment where an animal or plant normally lives and grows.

Heat—the time in a reproductive cycle when females are receptive to mating and can become pregnant; also called estrus.

Herdbook—a published record of animals which have been registered in a specific breed. Information is listed by registration number, and it includes name, pedigree and date of birth. The term herdbook is used for breeds of cattle, goats, and swine; studbook for horses and donkeys; and flockbook for breeds of sheep.

Heterosis—the performance boost obtained through the crossing of distantly related parents; also called hybrid vigor.

Husbandry—the care and management of animals. Intensive husbandry implies a high degree of intervention and use of many inputs, including high protein feeds, supplements, growth enhancers, and climate controlled housing. Extensive husbandry implies management with less intervention, with animals kept out of doors and harvesting their own feed in the form of forage.

Hybrid—the offspring of two parents of different breeds or species.

Hybrid vigor—the performance boost obtained through the crossing of distantly related parents; also called heterosis.

Improvement—the management and reproduction of a herd or flock of animals for the enhancement of specific production or conformation traits. Improvement can have as a goal the increase of uniformity in a population, particularly uniformity of performance, or the production of outstanding individuals with less regard for the uniformity of the population.

Inbreeding—the mating together of animals which are closely related. The resulting offspring have one or more ancestors on both the sire's side and the dam's side of the pedigree. Inbreeding increases the consistency and uniformity of a population, but can also reveal genetic weaknesses in a population.

Inbreeding coefficient—the mathematical calculation of the amount of inbreeding in individuals, which can be averaged for a population.

Industrial breeds—breeds which are specialized for rapid, maximum production of a single product (such as meat or milk) when given high levels of inputs (such as grain feed) and kept in a closely controlled environment. Individuals within industrial breeds tend to be genetically uniform and predictable.

Inputs—feed, equipment, housing, husbandry, medical care and other provisions given to animals in order to keep them healthy and productive.

Landrace—a livestock breed shaped through natural as well as human selection. Landraces demonstrate a high degree of environmental adaptation, fertility, and parasite and disease resistance, but show less phenotypic consistency than standardized breeds.

Linebreeding—the type of inbreeding which concentrates on one given ancestor with the goal of concentrating the genetic impact of this individual in the population.

Linecrossing—the mating of animals from different bloodlines within a single breed. Linecrossing can yield a performance boost in the first generation, somewhat like crossbreeding.

Livestock—domestic animals which are used in agriculture for the production of food or fiber, or used to provide agricultural services. Globally, many species are considered livestock, including camels, guinea pigs, elephants, llamas, rabbits, reindeer, and water buffalo. The traditional livestock species in North America are cattle, donkeys, goats, horses, sheep, and swine. The poultry species (chickens, turkeys, ducks, and geese) are also considered livestock in America.

Multiple sire mating—a system that uses multiple males at a single time to breed with a group of females. The result is that pedigrees are not easily determined.

Natural selection—the favoring of certain genetic characteristics in a population by the natural environment.

Open herdbook—a registry policy which allows the registration of individuals which are not the offspring of registered parents. Policy may require inspection and approval of individual animals prior to registration or may allow individuals or offspring of an approved breed to be registered. Depending on the specific mechanisms instituted, some open herdbook breeds are true breeds in the genetic sense, while others are not.

Outbred—the opposite of inbred, usually implying that an animal resulted from a recent "out" cross with an unrelated individual or bloodline.

Pedigree—a list, usually in chart form, of an individual's ancestors.

Performance—the demonstration of ability at a specific agricultural task or set of tasks, which varies among species. Production of beef or milk is the performance task for which cattle are measured, while the speed of a race horse is its relevant performance. Performance is controlled by genotype as well as by environment, including human management.

Phenotype—the outward appearance, behavior, and performance of an individual or breed, which is determined by its genotype and its environment.

Preservation—the protection of resources from loss through overexploitation, destruction, or neglect. The term implies keeping resources exactly as they were without allowing for change. Living animals are "preserved" through cryopreservation, which is the freezing of genetic materials, especially semen and embryos.

Primitive—animals or populations that are relatively unselected or unimproved. Primitive implies an animal which has been shaped more by natural selection than by human selection.

Randomized or random breeding—the management of a herd or flock to allow as many individuals as possible to mate and reproduce, with no deliberate pairing of animals. Theoretically this also involves minimal selection of parents, otherwise such systems become multiple sire mating systems. Randomized breeding contrasts to the specific mating of selected animals or groups of animals.

Relict traits—traits which persist in some populations as holdovers from the past history of the breed. This term is often used in regard to resistance traits which are largely gone from more improved breeds but persist in many of the rarer breeds.

Replacements—the next generation, especially the members of the next generation which represent the specific genetics of their parents.

Reproductive fitness—the ability to reproduce efficiently.

Resistance—the ability to thrive in the face of adverse environmental factors such as disease, parasites, poor quality feeds, or minimal management.

Ruminant—a cud-chewing mammal of the suborder Ruminantia. These mammals have complex four chambered stomachs that enable them to digest rough forage. Ruminants include cattle, goats, sheep, and deer.

Seasonal reproduction—the biological limitation of mating to certain times of the year, which is true in sheep, goats, horses, and several other species. This is usually a female effect, as females are only in estrus during the specific mating season. Some breed differences within a species may exist.

Selection—the act or process of choosing some individuals from out of a group based on fitness or preference, so that some individuals survive and propagate but others do not. As a result, the inherited traits of the survivors are perpetuated, while the traits of others are not.

Sire—the male parent of an animal.

Sound—healthy and free from defects.

Standardized breed—a breed of livestock or poultry shaped primarily by human selection, defined by a written breed description (or "standard"), documented through history, and promoted by an organized group of breeders. Most breeds of livestock and poultry in North America are considered standardized breeds.

Strain—a group of animals within a breed that are more closely related to each other than to the breed as a whole; also called bloodline.

Studbook—a published record of animals which have been registered in a specific breed. Information is listed by registration number, and it includes name, pedigree and date of birth. The term herdbook is used for breeds of cattle, goats, and swine; studbook for horses and donkeys; and flockbook for breeds of sheep.

Uniformity—the degree to which a breed, population, or species has minimal variation for each characteristic. This situation is often created by strong selection (human or natural) for a specific environment. A uniform population has less ability to adapt to changing conditions than does a variable population.

Variability—the degree to which a breed, population, or species has numerous genetic variants for each characteristic. This situation results from relatively relaxed selection (human or natural) for many traits. Genetically variable populations have more success in adapting to change than do uniform populations.

Viability—the capability for living. Viability may refer to an individual or to a group of animals. The viability of a group includes its ability to reproduce succeeding generations.

Selected Resources on Animal Husbandry and Breed Conservation

All Species

Books—General

Animal Agriculture: The Biology, Husbandry, and Use of Domestic Animals edited by H.H. Cole and W.N. Garret, 739 pp, W. H. Freeman & Co., 1980.

Animals in the American Economy by John Sims and Leslie Johnson, 228 pp, Iowa State University Press, 1972.

Applied Animal Nutrition: Feeds and Feeding, 504 pp, MacMillan Publishing Company, 1991.

A Chance to Survive: Rare Breeds in a Changing World, second edition, by Lawrence Alderson, 192 pp, Cameron and Tayleur, 1989, out of print.

Evolution of Domesticated Animals edited by I.L. Mason, 452 pp, Longman Group Ltd, 1984.

Farm Animal Behavior & Welfare by A.F. Fraser & D.M. Broom, 437 pp, Balliere Tindall, 1990.

Farm Animals in the Making of America by Paul C. Johnson, 128 pp, Wallace Homestead Book Co., 1975.

Fences for Pasture and Garden by Gail L. Damerow, 154 pp, Storey/Garden Way Publishing, 1992.

Grass Farmers by Allan Nation, 192 pp, Green Park Press, 1993.

Handbook of Livestock Management Techniques by Richard A. Battaglia and Vernon B. Mayrose, Prentice Hall Publishing Company, 1981.

Keeping Livestock Healthy (3rd edition) by N. Bruce Haynes, DVM, 344 pp, Storey/Garden Way Publishing, 1994.

Prescriptions for Healthy Farm Animals by John B. Herrick, 448 pp, Lessiter Publications (c/o No Till Farmer, Inc, PO Box 624, Brookfield, WI 53008-0624), 1991.

The Stockman's Handbook, seventh edition, by M.E. Ensminger, 1030 pp, Interstate Publishers Inc., 1992.

Taking Stock: The North American Livestock Census, by D.E. Bixby, C.J. Christman, C.J. Ehrman, and D.P. Sponenberg, 182 pp, McDonald and Woodward Publishing, 1994. (Available from the American Livestock Breeds Conservancy, PO Box 477, Pittsboro, NC 27312.)

Two Hundred Years of British Farm Livestock by Stephen J.G. Hall and Juliet Clutton-Brock, 272 pp, British Museum of Natural History, 1989.

A World Dictionary of Livestock Breeds, Types & Varieties, third edition, by I.L. Mason, 348 pp, Commonwealth Agricultural Bureaux, 1988.

Books—Genetics and Breeding

Animal Genetics by F.B. Hutt and B.A. Rasmussen, 582 pp, John Wiley and Sons, New York, 1982.

Bantam Breeding and Genetics by Fred P. Jeffrey, 308 pp, Saiga Publishing Co., 1977.

Breeding and Improvement of Farm Animals, sixth edition by V.A. Rice, F.N. Andrews, E.J. Warwick, and L.E. Legates, 477 pp, McGraw-Hill Book Co., New York. (This book is usually available at used book stores.)

Creative Poultry Breeding by Clive Carefoot, 220 pp, published by Carefoot (Sandy Bank, Chipping, Preston PR3 2GA England), 1985.

Genetics for the Animal Sciences by L.D. Van Vleck, E.J. Pollack, and E.A.B. Oltenacu, 391 pp, W.H. Freeman and Co., New York, 1987.

Genetics and Evolution of the Domestic Fowl by Lewis Stevens, 306 pp, Cambridge University Press, 1991.

Poultry Breeding and Genetics edited by R.D. Crawford, Elsevier (Amsterdam), 1990.

Poultry Breeding by A.L. Hagedoorn and Geoffrey Sykes, 240 pp, Crosby Lookwood and Son Ltd., 1953.

Veterinary Genetics by F.W. Nicholas, 580 pp, Oxford Science Publications, Oxford University Press, New Jersey, 1987.

Organizations and Periodicals

ATTRA—Appropriate Technology Transfer to Rural Areas, PO Box 3657, Fayetteville, AR 72702-3657, 800-346-9140, 501-442-9831. ATTRA provides information on sustainable agriculture to commercial producers.

The Cooperative Extension Service (or Agricultural Extension Service) has an office in almost every county in the United States with specialists on many agricultural enterprises and free information on animal husbandry. Consult your county government to find the nearest Cooperative Extension Service office.

Countryside and Small Stock Journal, N 26015 Winter Sports Rd, Withee, WI 54498, 800-551-5691.

Rural Delivery, Box 1509, Liverpool, Nova Scotia, B0T 1K0, Canada, 902-354-3321.

Rural Heritage, 281 Dean Ridge Lane, Gainesboro, TN 38562-5039.

Small Farm Today, 3903 W Ridge Trail Rd, Clark, MO 35243, 1-800-633-2535.

Small Farmers Journal, PO Box 1627, Sisters, OR 97759, 503-549-2064.

Small Scale Agriculture, OSSA, USDA-CSRS, Suite 342 Aerospace Building, Washington, DC 20250-2200, 202-401-4640.

Stockman Grass Farmer, PO Box 9607, Jackson, MS 39286-9607, 800-748-9808.

Successful Farming, 1716 Locust, Des Moines, IA 50309-3023, 1-800-374-3276.

Book Sellers

Acres USA Bookstore, PO Box 9547, Kansas City, MO 64133, 816-737-0064.

Ag Access, 603 4th St, Cavis, CA 95616, 916-756-7177.

Diamond Farm Book Publishers, Box 537, Alexandria Bay, NY 13607, 800-481-1353.

The Meadow Tree (horse books only), 3716 Effingham Place, Los Angeles, CA 90027.

The Mishka Farm Bookstore, N8246 Esterly Rd, White Water, WI 53190, 414-473-5595.

Storey/Garden Way Publishing, Schoolhouse Rd, Pownal, VT 05261, 1-800-451-3522.

Williamson Publishing, Box 185, Church Hill Road, Charlotte, VT 05445, 1-800-234-8791.

Cattle and Oxen

Basic Training of Oxen and *Advanced Training of Oxen* by Drew Conroy, Butler Publishing and Tools, PO Box 1390, Laporte, CO 80535, 800-728-3826.

Breeds of Cattle by Herman R. Purdy and R. John Dawes, 282 pp, Chanticleer Press, Inc., 1987.

Dairy & Cattle Science, third edition, by M.E. Ensminger, 550 pp, Interstate Publications, Inc., 1993.

The Family Cow by Dirk Van Loon, 262 pp, Storey/Garden Way Publishing, 1976.

Hoard's Dairyman magazine, PO Box 801, Fort Atkinson, WI 53538, 414-563-5551.

National Cattleman's Association, PO Box 3469, Englewood, CO 80155, 303-694-0305.

The Oxen Handbook by Drew Conroy and Dwight Barney, Butler Publishing and Tools, PO Box 1390, Lapote, CO 80535, 800-728-3826.

Purebred Dairy Cattle Association, c/o Bradford Ellsworth, PO Box 160, Cobol, MO 65689, 417-962-3141.

Tillers International, 5239 S 24th St, Kalamazoo, MI 49002. (Tillers provides information and instruction on the use of oxen.)

Training the Teamster by Leslie Barden, 12 pages, 1984, available from Leslie Barden, Barden Tree Farm, 357 Meaderboro Rd, Rochester, NH 03867.)

Donkeys

The Brayer magazine, American Donkey and Mule Society, 2901 N Elm St, Denton, TX 76201, 817-382-6845.

The Definitive Donkey by Paul and Betsy Hutchins, 221 pp, the American Donkey and Mule Society, 2901 N Elm St, Denton, TX 76201, 817-382-6845.

The Modern Mule by Paul and Betsy Hutchins, 136 pp, available from the American Donkey and Mule Society, 2901 N Elm St, Denton, TX 76201, 817-382-6845.

The Professional Handbook of the Donkey, second edition, by Elisabeth D Svendsen, MBE, 284 pp, The Donkey Sanctuary, 1986, available from the American Donkey and Mule Society.

Your Donkey by Ruth Harris, 96 pp, Broomtail Publications, 1994, available from the American Donkey and Mule Society.

Goats

American Dairy Goat Association, PO Box 865, Spindale, NC 28160, 704-286-3801.

American Meat Goat Association, PO Box 979, Mertzon, TX 76941.

Dairy Goat Journal, W2977 Markert Rd, Helenville, WI 53137, 414-593-8385.

Goats and Goat Keeping by Katie Thear, Mere Hurst Press, 1988.

The Pack Goat by John Mionczynski, Wind River Pack Goats, Rt 62, Box 250, Atlantic City, WY 82520.

Raising Milk Goats the Modern Way, revised edition, by Jerry Belanger, 208 pp, Story/Garden Way Publications, 1990.

Raising Milk Goats Successfully by Gail Luttman, 176 pp, Williamson Publishing, 1986.

Your Goats: A Kid's Guide to Raising and Showing by Gail Damerow, 172 pp, Storey/Garden Way Publishing, 1993.

Horses

The Complete Book of the Horse by Carol Foster, 208 pp, American Nature Society Press, Crown Publications Inc., 1983.

The Draft Horse Journal, Box 670, Waverly, IA 50677, 319-352-4046.

The Draft Horse Primer, second edition, by Maurice Telleen, 386 pp, *The Draft Horse Journal,* Box 670, Waverly, IA 50677, 319-352-4046.

Driving Digest Magazine, PO Box 467, Brooklyn, CT 06234, 203-546-6134.

The Encyclopedia of the Horse by Harley Edwards, 400 pp, Dorling Kindersley Publications, 1994.

Equus magazine, 656 Quince Orchard Rd, Gaithersburg, MD 20878, 301-977-3900.

Horse Industry Directory, annual publication of the American Horse Council, 1700 K Street, NW, Suite 300, Washington, DC 20006-3805, 202-296-4031.

Horsekeeping on a Small Acreage: Facilities Design and Management by Cherry Hill, 179 pp, Storey/Garden Way Publishing, 1990.

The Reach: The Driving World/Horse and Harness Directory (annual publication) PO Box 632, Kendallville IN 46565, 219-347-8223.

Training Workhorses, Training Teamsters by Lynn R. Miller, 352 pp, 1994, *Small Farmers Journal* PO Box 1627, Sisters, OR 97759, 503-549-2064.

The Work Horse Handbook by Lynn R Miller, 224 pp, *Small Farmers Journal*, PO Box 1627, Sisters, OR 97759, 503-549-2064.

Pigs

Fit for a Pig: Low Cost, Sustainable Strategies of Resourceful Hog Farmers by Shawn Grulla, 46 pp, ca. 1992, Center for Rural Affairs, Box 736, Huntingdon, NE 68739, 402-254-6893.

Raising Pigs Successfully by Kathy and Bob Kellog, 192pp, Williamson Publishing, 1985.

Small Scale Pig Raising, by Dirk Van Loon, 272 pp, Storey/Garden Way, 1978.

Poultry

American Standard of Perfection, published periodically by the American Poultry Association, 72 Springer Lane, New Cumberland, PA 17070, 717-774-1926. Most recent edition was published in 1993.

American Poultry History 1823-1973: An Anthology Overview edited by John L. Skinner, 775 pp, American Poultry Historical Society (Madison, WI) 1974. Available from Watt Publishing Co., 22 S. Wesley Ave, Mount Morris, IL 61504-1497.

American Poultry History 1973-1993 edited by John L. Skinner, 320 pp, Watt Publishing Co., 122 S. Wesley Ave, Mount Morris, IL 61504-1497, available in 1995.

The Book of Geese: A Complete Guide to Raising the Home Flock by Dave Holderread, 212 pp, The Hen House PO Box 492 Corvallis, OR 97730, 1981.

The Chicken Health Handbook by Gail Damerow, 344 pp, Storey/Garden Way Publishing, 1994.

Chicken Tractor: The Gardener's Guide to Happy Hens and Healthy Soil by Andy Lee, 248 pp, Good Earth Publications, 1994.

Chickens In Your Backyard: A Beginner's Guide by Gail and Rick Luttman, 157 pp, Rodale Press, 1976.

Domestic Ducks and Geese by Paul P. Ives, 372 pp, Orange Judd Publishing Company, 1947.

Ducks and Geese In Your Backyard: A Beginner's Guide by Gail and Rick Luttman, 273 pp, Rodale Press, 1978.

Modern Waterfowl Management by Oscar Grow, 358 pp, American Bantam Association, 1972.

Poultry Production by Richard Austic and Malden Nesheim, 325 pp, Lea and Febiger, 1990.

Raising the Home Duck Flock: A Complete Guide by Dave Holderread, 174 pp, Storey/Garden Way Publishing, 1978.

Raising Poultry the Modern Way, revised edition, by Leonard S. Mercia, 240 pp, Storey/Garden Way Publishing, 1990.

Raising Poultry Successfully by Will Graves, 192 pp, Williamson Publishing, 1985.

Raising Turkeys, Ducks, Geese, Pigeons, and Guineas by Cynthia Haynes, 354 pp, TAB Books, 1987.

Raising Your Own Turkeys by Leonard Mercia, 144 pp, Storey/Garden Way Publishing, 1981.

Your Chickens: A Kid's Guide to Raising and Showing by Gail Damerow, 156 pp, Storey/Garden Way Publishing, 1993.

Sheep

The American Sheep Industry Association, 6911 S. Yosemite St, Englewood, CO 80112-1414, 303-771-3500. ASIA has several publications, including *National Lamb and Wool Grower* magazine.

Black Sheep Newsletter, 25455 Dixie Mountain Rd, Scappoose, OR 97056.

Fiberfest (annual summer event), PO Box 46, Freeport, MI 49325.

The Marker, PO Box 487, Willits, CA 95490.

The Maryland Sheep and Wool Festival (annual spring event), PO Box 99, Glenwood, MD 21738.

The New Fiberworks Sourcebook: Your Essential Mail-order Guide to Supplies and Services for the Fiber Arts by Bobbie A. McRae, 1338 Ross St, Petaluma, CA 94954-6502, 800-BUY-YARN.

North American Dairy Sheep Association, Rt 3, Box 10, Hinckley, MN 55037.

Practical Sheep Dairying: The Care and Milking of the Dairy Ewe by Olivia Mills, 224 pp, Thorsons Publishers Ltd., 1982.

Raising Sheep the Modern Way by Paula Simmons, 278 pp, Storey/Garden Way Publishing, 1989.

sheep! magazine, W 2997 Markert Rd, Helenville, WI 53137, 414-593-8385.

Sheep & Wool Science, fourth edition, by M.E. Ensminger, 948 pp, Interstate Publishers, Inc,, 1970.

The Sheep Raiser's Manual by William K. Kruesi, 288 pp, Williamson Publishing, 1985.

The Shepherd magazine, 5696 Johnston Rd, New Washington, OH 44854, 419-492-2364.

Turning Wool into a Cottage Industry by Paula Simmons, 192 pp, Storey/Garden Way Publishing, 1991.

Your Sheep: A Kid's Guide to Raising and Showing by Paula Simmons and Darrell L. Salsbury, 128 pp, Storey/Garden Way Publishing, 1993.

Photography Credits

7, Piney Woods cow and calf, D.P. Sponenberg
8, American Mammoth Jack, Brigitte Lange
9, White Leghorn chickens, U.S.D.A.
10, Santa Cruz, Joe McCummins
12, Myotonic goats, D.P. Sponenberg
13, Dutch Belted cattle, Paul Daniels
14, Jacob sheep, Ingrid Painter
18, Navajo-Churro sheep, Lyle McNeal
20, Bronze turkeys, David Sullenberger
22, Gulf Coast Native sheep, Don Bixby
24, Tamworth pigs, Don Bixby
26, Milking Shorthorn cow, Don Bixby
27, Randall Lineback calf, D.P. Sponenberg
28, Commercial cattle, U.S.D.A.
29, Suffolk horses, Maryanne Mott
31, White Rock chicken, Don Bixby
36, Movable chicken house, Don Bixby
37, Tamworth pigs, Don Bixby
39, Galloway cattle, Evelyn Simak
40, Piney Woods oxen, D.P. Sponenberg
41, Morgan horses, Margaret Gardiner
42, Horse with rider, Sam Getzen
43, Mule with rider, Kathy Suits-Smith
44, Jennet and foal, D.P. Sponenberg
45, Tunis sheep, Rob Amberg
46, Katahdin sheep, Piel Farm
47, Leicester Longwool sheep, Kathy Suits-Smith
48, Goat cart, courtesy of Don Bixby
49, San Clemente goats on rocks, Sheri Tepper
50, Gloucester Old Spot pigs, Don Bixby
51, Mulefoot pigs, Mark Fields
52, Girl with New Hampshire rooster, Susanna Davy
53, Dominique chickens, Horne Creek Farm
54, Eggs, Don Bixby
55, Khaki Campbell drake, Heather Maxey
56, Pilgrim gander, Rob Amberg
57, Bronze turkey, David Sullenberger
60, Couple with Jacob sheep, courtesy of W. and E. Van Valkenberg
61, Highland cow, Gordon Kohl
65, Man holding Tamworth piglets, Gene Hebert
68, Women with Dutch Belted and Dexter heifers, courtesy of R. Fleharty
72, Florida Cracker cow, courtesy of T.A. Olson
73, Florida Cracker cow with man, D.P. Sponenberg
77, Milking Devons on milk line, Robert Gear
80, Gloucester Old Spot pigs, Elizabeth Henson
82, White Park cattle, Don Bixby
84, Dexter cow and calf, Maryanne Mott
87, St. Croix sheep, Stephan Wildeus
91, Myotonic goats, D.P. Sponenberg
92, Spanish Mustang herd, D.P. Sponenberg
93, American Cream foal, courtesy of Elaine Shirley
94, Karakul ewe and lamb, Deborah Simon
95, Shetland sheep, Linda Zuppan
98, Man with Leicester Longwool lamb, Elaine Shirley
99, Lineback cattle, D.P. Sponenberg
100, Gulf Coast Native sheep, Tom Gray
101, Spanish Mustang foal, Robin Keller
106, New Hampshire rooster, Rob Amberg
107, Delaware chickens, Carolyn Christman
108, Bronze turkey poults, David Sullenberger
110, Dominique fledgling, Roger Boyles
111, New Hampshire chicks, Rob Amberg
114, Pilgrim geese, Don Bixby
116, Woman with Light Brahma chicken, courtesy of Julie Rawson
117, Saxony Indian Runners, Heather Maxey
120, Man with Exmoor pony, Don Bixby

About the Authors

D. Phillip Sponenberg is a Professor of Pathology and Genetics at the Virginia-Maryland Regional College of Veterinary Medicine at the Virginia Polytechnic Institute and State University located in Blacksburg, Virginia. He received a D.V.M. from Texas A & M University and a Ph.D. from Cornell University. Dr. Sponenberg is Technical Coordinator for the American Livestock Breeds Conservancy and has written widely on issues of livestock genetic conservation and animal genetics.

Carolyn J. Christman is the Program Coordinator of the American Livestock Breeds Conservancy in Pittsboro, North Carolina. An historian and former high school teacher, she received an A.B. in History from the University of North Carolina and an M.A. in Anthropology from Wake Forest University. She has co-authored several publications on rare breed conservation, including *Taking Stock: The North American Livestock Census.*

Index